INTERNATIONAL MINING FORUM 2006

BALKEMA – Proceedings and Monographs
in Engineering, Water and Earth Sciences

New Technological Solutions in Underground Mining
International Mining Forum 2006

Eugeniusz J. Sobczyk
Polish Academy of Sciences, Mineral and Energy Economy Research Institute, Cracow, Poland

Jerzy Kicki
AGH – University of Science and Technology, Department of Underground Mining, Cracow, Poland
Polish Academy of Sciences, Mineral and Energy Economy Research Institute, Cracow, Poland

Taylor & Francis
Taylor & Francis Group

LONDON / LEIDEN / NEW YORK / PHILADELPHIA / SINGAPORE

Library of Congress Cataloging-in-Publication Data
Applied for

Published by: Taylor & Francis / Balkema
                  P.O. Box 447, 2300 AK Leiden, The Netherlands
                  e-mail: Pub.NL@tandf.co.uk
                  www.balkema.nl, www.tandf.co.uk, www.crcpress.com
ISBN10 0-415-40117-8
ISBN13 978-0-415-40117-3

Printed in EU

International Mining Forum 2006, Sobczyk & Kicki (eds) © 2006 Taylor & Francis Group, London, ISBN 0415 401178

Table of Contents

International Mining Forum 2006, Sobczyk & Kicki (eds) © 2006 Taylor & Francis Group, London, ISBN 0415 401178

Preface

Professor Boleslaw Krupinski, an enlightened miner, the organizer and first Chairman of the International Organizing Committee of the World Mining Congress so spoke about the mining profession: "A miner always was, is and will be a man of technical and social progress. Only through the perfecting of social and technical work conditions, can a miner discover the secrets and treasures of the Earth, conquer and exploit them for the benefit of all, reverse dangers caused by nature and provide the country with forces of nature".

These beautiful words are the motto of the School of Underground Mining, which is the organizer of the International Mining Forum – a meeting of practitioners and scientists from many countries. Following these words we try to include in the Forums' proceedings papers addressing a broad range of subjects, many a time problems extending beyond underground mining.
So it is this time.

Along with topics strictly connected with underground mining, e.g.:

– techniques and technologies in underground mines,
– rock engineering problems,
– monitoring of natural hazards in underground coal mines,

we present papers important to the development of the mining industry, related to:

– bringing into being the idea of sustainable development,
– limiting the greenhouse effect.

The International Mining Forum was held thanks to the support of the Chair of Underground Mining, the Faculty of Mining and Rock Engineering of the University of Science and Technology (AGH), the Mineral and Energy Economy Research Institute of the Polish Academy of Science in Cracow, KGHM Polska Miedz S.A., FTT Stomil Wolbrom, Katowice Coal Holding S.A., LW Bogdanka S.A., Elgór+Hansen S.A., MIDO Ltd., ZOK Ltd.

The organizers would also like to express their gratitude to all other persons, companies and institutions, who helped in bringing the Forum into being.

We hope that the Forum will contribute to the exchange of interesting experiences and establishing new acquaintances and friendships.

Jerzy Kicki
Chairman of the Organizing Committee 2006

Organization

Organizing Committee:
Jerzy Kicki (Chairman)
Eugeniusz J. Sobczyk (Secretary General)
Artur Dyczko
Jacek Jarosz
Piotr Saługa
Krzysztof Stachurski

Advisory Group:
Prof. Volodymyr I. Bondarenko (National Mining University, Ukraine)
Mr. Wojciech Bradecki (State Mining Authority, Poland) – Chairman of IMF 2005
Prof. Jan Butra (CUPRUM Ltd., Poland)
Dr. Alfonso R. Carvajal (Universidad de La Serena, Chile)
Prof. Piotr Czaja (AGH – University of Science and Technology, Poland)
Prof. Bernard Drzęźla (Silesian University of Technology, Poland)
Prof. Józef Dubiński (Central Mining Institute, Poland)
Prof. Jaroslav Dvořáček (Technical University VSB, Czech Republic)
Prof. Paweł Krzystolik (Experimental Mine Barbara, Poland)
Prof. Garry G. Litvinsky (Donbass State Technical University, Ukraine)
Prof. Eugeniusz Mokrzycki (Polish Academy of Sciences, MEERI, Poland)
Prof. Roman Ney (Polish Academy of Sciences, MEERI, Poland)
Prof. Jacek Paraszczak (University of Laval, Canada)
Prof. Janusz Roszkowski (AGH – University of Science and Technology, Poland)
Prof. Stanisław Speczik, (Polish Geological Institute, Poland)
Prof. Anton Sroka (Technische Universität Bergakademie Freiberg, Germany)
Prof. Mladen Stjepanovic (University of Belgrade, Yugoslavia)
Prof. Antoni Tajduś (AGH – University of Science and Technology, Poland)
Prof. Kot F. v. Unrug (University of Kentucky, USA)
Dr. Leszek Wojno (Australia)
Dr Yuan Shujie (Anhui University of Science and Technology. Huainan, Anhui, Republic of China)

International Mining Forum 2006, Sobczyk & Kicki (eds) © 2006 Taylor & Francis Group, London, ISBN 0415 401178

Coal Mine of 21st Century:
In-Situ Producer of Energy, Fuels and Chemicals

Józef Dubiński, Jan Rogut, Krystyna Czaplicka, Aleksandra Tokarz
Główny Instytut Gornictwa. Katowice, Poland

ABSTRACT: This discussion paper presents some innovative, balanced visions and scenarios of new types of a coal mine development. Such a novel coal mine as an in-situ producer of energy, liquid fuels and chemicals could satisfy the increasing needs of European and Polish economy at the time of expensive energy. The vision outlines a response of coal mining research community to the requirements of energy and chemical industries in order to continue a secure supply of energy carriers, fuels and chemicals at the post "peak oil" era. The paper puts under discussion the similarities and differences in thermal and chemical coal conversion processes carried out in-situ (underground conversion of coal) and out of the coal bed (Energyplex). The paper discusses selected needs for R&D activities aimed at achieving some critical, scientific and technical mass and knowhow necessary for the on time delivery of innovative technology solutions for sustainable coal mining industries. The paper also accentuates the need for elaborating extended LCA techniques for the economic and environmental evaluation of different options of the future coal mine structure and functions. The text draws out several recommendations for the long-term coal conversion technology strategies in line with Strategic Research and Strategic Deployment Agendas (SRA, SDA) of Zero Emission Fossil Fuel Power Plant, Hydrogen & Fuel Cell, Green Chemistry and similar European Technology Platforms as well as the European Sustainable Mining and Mineral Extraction Technology Platform.

1. THE IDEA

The backbone of this paper consists in "a return to the future" approach, due to the fact, that the ideas presented here as prospective for the period of 2010–2030, are based substantially on updated experiences and expertise from in-situ coal processing that is underground coal gasification tests taking place in several locations in the middle of 20th Century combined with the results of Fisher-Tropsch liquid fuel synthesis developed in the II World War in Germany.

The most recent renaissance of interest in Clean Coal Conversion Technologies in Europe is based on the growing demand of energy and fuels per capita in the situation when the access to the low cost energy resources continuously decreases. This generates the necessity of return in the medium or long term scale to exploitation of the deposits of fossils treated previously as non economic or exhausted (secondary mining). Therefore, some radical but rational rethinking of the targets and directions of R&D programs of mining sciences. The time has come to realize that easily available and economic coal deposits have been exhausted and efforts are necessary to remedy such a situation. A sound example would be here the technology development of a highly productive coal extraction method using long walls. Those mature systems of high efficiency have been developed for the extraction of large and thick regular deposits of coals, on surface or underground.

The situation is rapidly changing, as (excluding some locations in US) the availability of rich deposits of coal with favourable geometry of seams decreases fast, which means that the long term R&D preferences for coal extraction industry need to be re-targeted based on a possibly precise foresight (TOR – terms of reference) of the future development of global energy technology.

In this paper, the following statements have been chosen as the background of analysis:

1. Solid fossil fuels will be available for the production of energy, fuels and chemicals all through the 21st century; however, coal exploitation will be continuously transferred from the extraction of large and rich deposits (thick, regular surface or moderately deep) to the thin and irregular ones requiring sophisticated intelligent technologies.

2. The portfolio of acceptable coal extraction technologies will be continuously reduced due to the growing concerns of environmental regulations and risk assessment of natural and human related hazards.

3. The preferences will be growing for integral resource utilization, recycling of products and processing fluids, the extension of the service life of tools for preparatory and extraction works, saving the energy and intelligent recognition and use of the deposits.

4. The decision makers will be pressed to utilize the life cycle techniques in the evaluation of the optimal technology choice.

2. IN-SITU COAL CONVERSION – ISCC PROCESS

The basic concepts of the innovative in-situ coal conversion will be illustrated here by several case studies recalled from the most recent patents, PhD and MSc theses and review papers aimed at different issues related to in-situ coal mining and processing. The examples do not exhaust the existing opportunities and were chosen for illustrative purposes of the main line of this paper.

2.1. In-situ thermal processing of coal

In-situ thermal processing of coal as an innovative idea is protected by numerous most recent US and global patents issued to Royal Dutch Shell (Houston, TX). Figure 1a (adopted from the US Patent 6866097 – March 15, 2005) illustrates the yields of different products which could be generated from coal deposits in-situ at several stages of coal formation heating (pyrolysis). Y axis descrybes yield as barrels of oil equivalent per ton versus temperature on x axis. The drawing is the mirror picture to the one known from thermal conversion of biomasses (Fig. 1b).

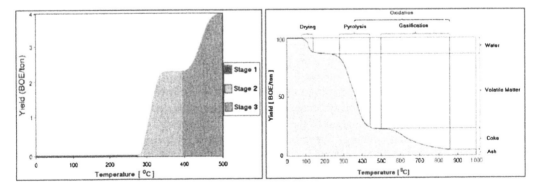

Figure 1a (left). Several stages of heating a coal formation
Figure 1b (right). Biomass thermal conversion

According to the basic concept of Shell, the temperature necessary for supporting the process of mild degasification and thermolysis of coal seam is about 200°C or 250°C (as shown in Figure 1a, stage 1). The source of energy for the bed heating is a controlled combustion of part of it. An oxidant necessary for the process (air, oxygen, or oxygen enriched air) may be introduced into a portion of the formation with synthesis gas generating fluid. The oxidant may react with carbon in the formation to produce CO_2 and/or CO. At specified conditions the oxidation process could be carried out according to the flame-less mode. In this stage, desorbed methane and vaporized water may be produced from the formation. Mild thermolysis of coal (through the external heating of coal bed) is seen as a potentially effective method of coal bed pre-treatment (degasification). The optimal temperature within a heated coal formation depends on heat sources, thermal conductivity and thermal diffusivity of the formation, the type of coal formation and composition of the hydrocarbons in the coal formation, water content of the formation, and/or type and amount of the oxidant.

The coal seam, with the internal structure modified by mild degasification, may be heated further to the pyrolysis temperature. The process consists in thermal non-oxic degradation of the substance during the long time heating. The pyrolysis temperature for major types of coal deposits ranges from approximately 250°C to 900°C. In stage 2, the temperature of the formation increases, the amount of condensable hydrocarbons in the produced formation fluid tends to decrease, and the formation will produce mostly methane and hydrogen.

The unused coal formation can be further gasified for an in-situ generation of synthesis gas. Synthesis gas is a mixture containing mainly hydrogen and carbon monoxide and water, carbon dioxide, nitrogen, methane and other gases. Synthesis gas may be produced within a temperature range from about 400°C, to about 1200°C during stage 3. Synthesis gas production may be commenced after the production of pyrolysis fluids has been substantially exhausted or becomes uneconomical.

2.2. The in-situ coal conversion on the Van Krevelen diagram

A systematic analysis of the relationships between the temperature and productivity of a heated coal bed could be carried out using the well known in coal chemistry Van Krevelen diagram, locating the varieties of carbonaceous solids in the H/C (hydrogen to carbon) and O/C (oxygen to carbon) space. Figure 2a shows this part of the diagram, which relates to the natural conditions taking place in the underground deposits of coal. For this paper, an extended diagram adopted from [Tepper 2005], thesis has been found more useful.

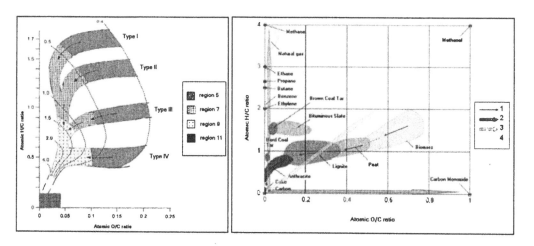

Figure 2a (left) and 2b (right). Maturation sequence for various types of kerogens

3

Figure 2 Van Krevelen diagram for coal conversion processes (Figure 2a) presents the standard coal bed conditions, thermal pyrolysis in-situ, and the extended Van Krevelen diagram illustrates the coal gasification (horizontal arrow 2), coal hydrogasification (vertical arrow 3) and complex chemical conversion (diagonal arrow 4). The diagram presented in Figure 2a shows the maturation sequence for various types of kerogens under temperature, pressure, and biochemical degradation for the natural (geologic time scale) conditions of biomass conversion. Figure 2b gives a broader picture of O/H, O/C space and could be employed for a more detailed explanation of the targeted in-situ coal conversion processes [4].

The natural conditions at coal deposits rarely extend the area presented in Figure 2a (the square inside the 0.0–0.4, 0–2 area of O/H and C/H diagram in the Figure 2b).

However, what we are really interested to do during the in-situ processing of coal are the following:
– Targeting the point (1.0, 0.0) – to produce carbon monoxide, transferred next to hydrogen and carbon dioxide.
– Aiming at the point (0.0, 4.0) – for in-situ generation of methane gas, the potential source of synthetic natural gas to be introduced into the existing natural gas pipelines.
– Focusing on the production of synthesis gas (mixture of carbon monoxide and hydrogen).
– For making the second approach applicable, the necessary amount of hydrogen gas must be generated in-situ or out of the bed and combined into the methane or other hydrocarbon products of interests, as shown in H/C line.

However, it is rather difficult to expect the possibility of approaching the production of specialty chemicals in-situ in the direct way – for example to produce methanol of DME.

So far, several demonstration scale and mature ex-situ technologies exist and they are able to produce hydrogen, methane and synthesis gas through thermal, oxythermal and hydrothermal conversion of carbon containing materials (coke, coal, biomass and bio-waste). There is a shortage of similar technologies, which could be used for a comparable conversion of coal (kerogens) onsite. The most recent patents and papers related to in-situ coal conversions suggest that serious, systematic and scientifically sound exploration of the assets and problems related to the subject have only started (see the citations part).

2.3. In-situ physical and chemical coal conversion

The major difference in the recent developments in the in-situ coal conversion research versus the old one is the multidisciplinary character of the efforts.

2.3.1. Formation of the permeability of the bed

For the precise control of the in-situ coal processing to gaseous or liquid products, the methods and tools for formation of communication system (building the bed permeability, for the introduction of the gasification agents, and for the recovery of the gasification products are of highest importance. It has been demonstrated by Shell that heating a portion of a coal formation in-situ to a temperature lower than an upper pyrolysis temperature may increase the permeability of the heated portion. This is due to the formation of the system of macropores during the evacuation of gaseous and liquid products from the heated coal bed. The experimental data recalled in the patent description demonstrate that fluids flow more easily through the heated portion. The other mechanism responsible for the permeability increase in heated coal beds is the creation of fractures resulting from thermal stresses. The fractures may also be generated due to fluid pressure increase as the result of coal thermal decomposition. The communication space generated during preliminary thermal degasification of coal seams depends on the type of coal (see Figure 3 and Table 1) and on the way the process is carried out.

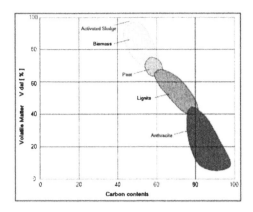

Figure 3. The volatile components contents versus carbon content for different carbonaceous raw energy carriers as the background for evaluation of the porosity level of the bed after mild pyrolysis [Tepper 2005]

Table 1. Properties of virgin coal and coal after thermal processing

	Deep Coal Formation	Post Treatment Coal Formation
Coal Thickness (m)	9	9
Coal Depth (m)	990	460
Initial Pressure (bars abs.)	114	2
Initial Temperature	25°C	25°C
Permeability (md)	5.5 (horizontal) 0 (vertical)	10,000 (horizontal) 0 (vertical)
Cleat porosity	0.2%	40%

2.3.2. In-situ chemical coal conversion

The Boudart reaction is a reaction of carbon dioxide with solid carbon generating carbon monoxide (left diagram) and the process of hydrogen production by CO conversion with water (normal pressure, equimolar starting composition). Both processes are strongly competitive under the bed conditions; however, their rational arrangement should generate a wide range of opportunities for process improvement. The on-site capture of carbon dioxide by different means seems to be one of the potential ways to increase the process efficiency.

The permeable bed of preliminary fractured coal is an attractive candidate for in-situ gasification and hydro gasification processes of coal to the mixtures of hydrogen, methane (synthetic natural gas) and synthesis gas.

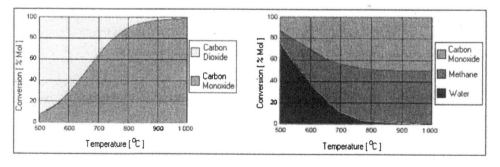

Figure 4a (left) and Figure 4b (right). Thermodynamics for Boudart reaction

The basic reactions involved in the formation of synthesis gas are as follows:

$C+H_2O->H_2+CO$,
$C+2H_2O->2H_2+CO_2$,
$C+CO_2->2CO$.

Thermodynamics allows the following reactions to proceed:

$2C+2H_2O->CH_4+CO_2$,
$C+2H_2->CH_4$.

In the presence of oxygen, the direct coal oxidation reaction may take place to generate carbon dioxide and heat:

$C+O_2->CO_2$.

2.3.3. *The Fisher-Tropsch process*

The fluid produced from a coal formation by an in-situ conversion process may include carbon monoxide that next may be used as feedstock for the Fischer-Tropsch process for the production of methanol or synthetic liquid fuels. The process is a catalysed chemical reaction in which carbon dioxide, carbon monoxide and methane are converted into liquid hydrocarbons of various forms, see the formula below.

$M+CO+H_2->M-CH_3+CO+H_2->M-C_nH_{2n+1}$,
$(n+2)CO+(2n+5)H_2->CH_3(--CH_2--)nCH_3+(n+2)H_2O$.

Hydrogen to carbon monoxide ratio for synthesis gas used as a feed gas for a Fischer-Tropsch reaction may be about 2:1.

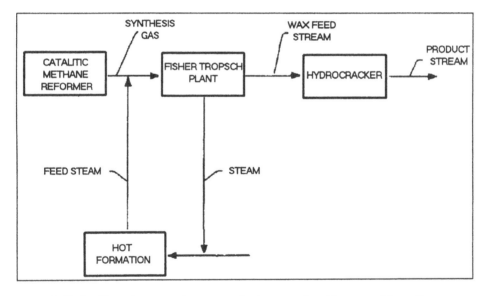

Figure 5. Fischer-Tropsch process that uses synthesis gas produced from a coal formation

Figure 5 illustrates the Fischer-Tropsch process that uses synthesis gas produced from a coal formation as a feed stream. Operating synthesis production wells at approximately 700°C may produce the proper ratio. The synthesis gas produced may also be used as a feed for a process of methanol production as well as the conversion of synthesis gas to gasoline and for a process that may convert synthesis gas to diesel fuel. Produced synthesis gas may be used as a feed gas for the production of ammonia and urea.

A series of experiments was conducted by Shell to determine the effects of various properties of coal formations on the properties of fluids produced from such coal formations. The series of experiments were conducted on forty-five cubes of coals to determine source rock properties of each coal and to assess potential oil and gas production from each coal. Figure 6 shows a comparison of gas compositions, in percentage volumes, obtained from in-situ gasification of coal using air or oxygen injection to heat the coal, and in-situ gasification of coal in a reducing atmosphere by thermal pyrolysis heating as described in embodiments herein.

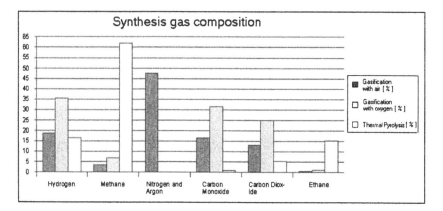

Figure 6. Comparison of compositions of product gas generated in-situ in gasification tests by Shell

Methods of producing synthesis gas were successfully demonstrated by Shell at the experimental field test. These included continuous injection of steam and air, steam and oxygen, water and air, water and oxygen, steam, air and carbon dioxide. All these injections successfully generated synthesis gas in the hot coke formation.

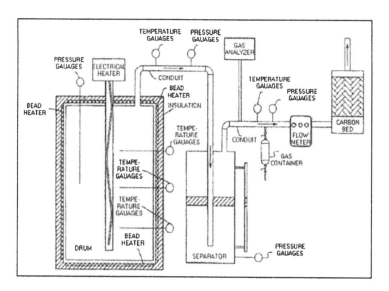

Figure 7. Experimental apparatus used in the Shell studies

The coal located in a drum was heated at a rate of about 2°C/day at various pressures. Compositions of the portion of condensable product fluids collected were determined by external analysis methods. Non-condensable fluids produced from pyrolysis of the cube and the drum such as hydrogen, methane, ethane, propane, n-butane had similar concentrations of various hydrocarbons generated within the coal. Crushing of the coal did not have an apparent effect on the pyrolysis of the coal.

3. THE IN-SITU CHEMICAL CONVERSION OF COAL AS THE TARGET FOR MULTIDISCIPLINARY RESEARCH

3.1. *Rational selection of the bed*

The coal formations for in-situ treatment should be selected taking into account not only the properties of carbonaceous solids, but also the economic, socio-economic and environmental criteria [1] as follows:
- Coal seam above 2 m thick – to be sure that the scale of productivity will cover the expenses of building the processing infrastructure.
- Depth between 600 and 1200 m with precise knowledge of the properties of the covering strata.
- The availability of good density borehole data.
- Stand off of above 500 m from abandoned mine workings, license areas.
- Above 100 m vertical separation from major aquifers.

3.2. *Coal properties after pyrolysis*

The heating of the coal bed in-situ in fact imitates the natural way of coal maturation (arrows 1 in Figure 2b). The major difference consists in the generation of the transport structure of the bed in the artificial (against the natural) process. Figure 8 illustrates the results from the analysis of the coal before and after it was treated.

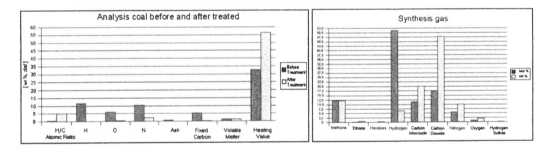

Figure 8 (left). Results from analysing the coal before and after it was treated
Figure 9 (right). Composition of synthesis gas produced in the in-situ coal field experiment

Interestingly, there appears the opportunity of radical improvement of the gasification process through an in-situ irreversible capture of CO_2 in part of the reactive surrounding strata or by the coal bed.

3.3. *The energy efficiency of the in-situ coal gasification*

The amount of useable energy contained in the produced synthesis gas is less than that contained in the pyrolysis fluids. It results from the overall thermodynamic scheme in which the synthesis gas production is less energy efficient than pyrolysis. Simply, the endothermic synthesis gas reaction consumes energy.

However, the Shell's data on the total energy efficiency of the in-situ coal conversion are quite optimistic. They reveal that the total energy content of the gasification products in the worst case is at least six times higher than the total energy consumed for heating the bed and for performing all the mechanical operations.

3.4. *The hydro-gasification process*

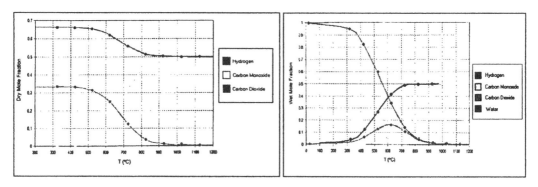

Figure 10 (left). Plot of calculated equilibrium gas dry mole fractions for a coal reaction with water
Figure 11 (right). Plot of calculated equilibrium wet mole fractions for a coal reaction with water

Figure 10 and Figure 11 adopted from an original Shell patent document illustrate that the lower end of the temperature range at which synthesis gas may be produced is about 400°C. At temperatures higher than about 660°C, the production of carbon monoxide over carbon dioxide is favoured.

Figure 12 shows the Shell's estimations for the potential production of different chemical commodities from the synthesis gas produced by in-situ coal wet oxidation. Their estimates are based on 56.6 million standard cubic meters of synthesis gas produced per day at 700°C.

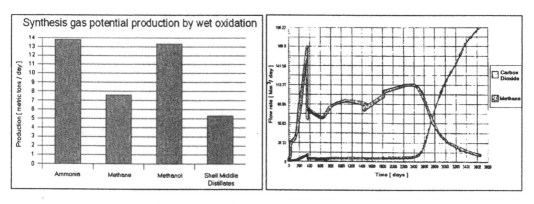

Figure 12 (left). Potential of different market products from the synthesis gas
Figure 13 (right). Shell's data showing that the majority of the injected carbon dioxide
is being sequestered in the coal formation. The data correlate well with the "Recopol" tests results

3.5. *Integrated capture and storage of carbon dioxide in processed coal deposits*

Carbon dioxide sequestration is the process of storing the by-product gas rather than venting it to the atmosphere. Higher quantities of carbon dioxide may be adsorbed in a coal formation at lower temperatures. The formation may be cooled by introducing water into it. The steam produced may be removed from the formation and used for any desired process. For example, the steam may be provided to an adjacent portion of the formation to heat it or to generate synthesis gas. Carbon dioxide may be sequestered in a deep coal formation as well as in a post treatment coal formation.

3.6. *The enhanced coal bed methane production*

The simulation model accounts for the matrix and dual porosity nature of coal and post treatment coal [3]. The spaces between the blocks are called "cleats" whereas cleat porosity is a measure of available space for the flow of fluids in the formation. Carbon dioxide and methane are assumed to have the same relative permeability. The cleat system of the deep coal formation was modelled as initially saturated with water because high level of water saturation inhibits the absorption of carbon dioxide within cleats. Therefore, water is removed from the formation before injecting carbon dioxide. The gases within the cleats may adsorb in the coal matrix. The matrix porosity is a measure of the space available for fluids to adsorb in the matrix.

The above processes are illustrated by original Shell data. After about 2400 days, the production rate of carbon dioxide started to rise significantly due to the onset of saturation of the coal formation. Methane was desorbing as carbon dioxide was adsorbing in the coal formation. The production rate of methane started to decrease after about 2400 days.

The innovative way of combining together the ex-situ coal conversion to hydrogen and electricity combined with enhanced methane recovery and carbon dioxide storage in coal bed is presented in Figure 14.

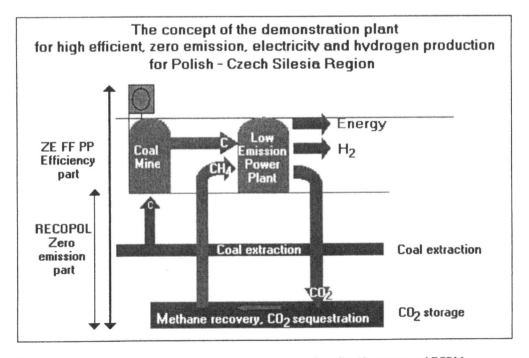

Figure 14. Integration of ex-situ coal conversion with in-situ carbon dioxide storage and ECBM

3.7. In-situ extraction of geothermal energy

Another innovative idea of using carbon dioxide under high pressure to extract energy from dry geothermal reservoirs has occurred. Namely, supercritical carbon dioxide is pumped into a hot formation and used to transfer the geothermal energy. On the surface, power generating plant fluid is recovered and sent back down into the reservoir. The heat-extraction process is repeated. Carbon dioxide is useful as the supercritical fluid because it is locally available, economical and easy to store and handle when not in contact with water. Also, it is inert, non-toxic and non-flammable for the environment.

3.8. In-situ conversion of coal to hydrogen and synthesis gas integrated with induced geothermy and carbon dioxide sequestration

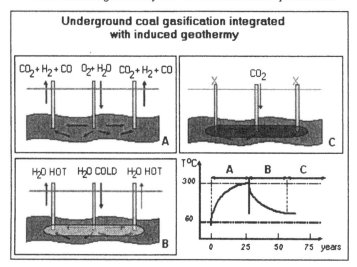

Figure 15. Integration of in-situ coal conversion with induced geothermy and carbon dioxide storage

The process of thermal conversion of coal in-situ when combined with further coal gasification and hydro-gasification will disperse in the surrounding strata a substantial amount of heat. This will result in a noticeable temperature increase of the underground processing space, high enough for the potential use as the source of geothermal energy. An important part of this energy could be recovered employing the standard geothermal heat recovery practice. Moreover, the cooled, post-coal conversion space could potentially serve as a stable storage vessel for carbon dioxide. Te above idea needs to be further elaborated, and some rational techniques for the evaluation of the long time economic and environmental risks and benefits must be adopted.

4. THE MAJOR RESEARCH NEEDS AND TECHNOLOGY ROADMAP FOR IN-SITU COAL CONVERSION PROCESSES

Responsible utilization of the great innovative potential of the in-situ coal conversion processes requires the start up of an integrated research and development program combining the following research areas:
1. Coal chemistry and physics examining the best process conversion options tailored to the properties of the available coal deposits.

2. Geology, hydrogeology and geophysics recognize the optimal location of ISCC units and systems taking into account the processing needs and the long time hazards and risk assessment and management.
3. Chemical and process engineering evaluating the optimal coal processing configurations combined with the efficient conversion, separation and purification of products and processing media.
4. Coal mining exploring the most effective ways of the ISCC system design, construction, implementation and exploitation.
5. Information technology tools and methods modelling and predicting the long-term performance of the underground system as an extremely complex in-situ chemical reactor.
6. Environmental, economic and socio-economic sciences supporting the decision makers in rational evaluation of the different technology options using the life cycle assessment approach.
7. Political and communication sciences evaluating the arguments and conditions for raising the awareness of local and regional communities in support of the ISCC technologies.

The elaboration of the technology roadmaps should start up from the most current review of the state of the art of coal conversion processes ex-situ and in-situ. This, in turn, will require rational extensions of the approach to cover and include arguments based on the data from the energy technology foresight exercises [5–8].

5. SUMMARY

5.1. *Main advantages of the in-situ coal conversion process (ISCC)*

The main advantages of the in-situ coal conversion process (ISCC) are:
– Minimal surface disturbance resulting from operations.
– Increased worker safety.
– No surface disposal of ash and coal tailings from coal washing plants.
– Minimal site rehabilitation.
– Reduced carbon dioxide emissions due to combined cycle efficiencies.
– Potential to remove CO_2 from the product gas before combustion, sequester it in the coal seam, and further reduce emissions.
– Practically unlimited supply of coal available for gasification, no coal and water supply is required to sustain the reaction.
– The ISCC process creates underground gas and heat storage capacity, which makes the gas supply very stable and robust.
– Absence of ash or slag removal and handling not necessary since they are left behind in the underground cavities.
– Ground water influx into the gasifier creates an effective "steam jacket" around the reactor making the heat loss in-situ tolerably small.
– Optimal pressure in the underground gasifier promotes ground water flow into the cavity, thus confining the chemical process to the limits of the gasifier and preventing contamination of the area.
– Supercritical carbon dioxide used as heat transfer in geothermal energy [2].

5.2. *The main challenges of ISCC*

The main challenges of ISCC consist in:
– Adaptation and optimisation of the well linking techniques.
– Adjustment of drilling techniques and completion methods.
– The necessity to operate the coal converter at minimal capacity to protect the resource while maintaining consistent quality of gas production.

- Detailed knowledge of proven geologic and hydro geologic conditions necessary for reliable design of the coal conversion facility with the extension to post extraction monitoring and control of potential hazards.
- Elaboration of reliable methods to predict and prevent the un-controlled emission of process media to the surrounding aquifers and protected areas [2].

REFERENCES

Review of the Feasibility of Underground Coal Gasification in the UK. DTI Report, September, 2004.

Walker L.K., Blinderman M.S., Brun K. 2001: An IGCC Project at Chinchilla, Australia Based on Underground Coal Gasification (UCG). Paper to 2001 Gasification Technologies Conference, San Francisco, October 8–10.2001.

Schoeling L., McGovern M. 2000: Pilot Test Demonstrates How Carbon Dioxide Enhances Coal Bed Methane Recovery. Petroleum Technology Digest, September 2000, pp. 14–15.

Tepper H. 2005: Zur Vergasung von Rest- und Abfallholz in Wirbelschichtreaktoren für Dezentrale Energieversorgungsanlagen. Dissertation, Magdeburg, Otto von Guericke University, Fakultät für Verfahrens- und Systemtechnik.

Praca zbiorowa pod redakcją K. Czaplickiej pt. Zastosowanie oceny cyklu życia (LCA) w ekobilansie kopalni. Wydawnictwo GIG, Katowice 2002 (ISBN 83-87610-48-8).

Czaplicka-Kolarz K. i inni: Model ekologicznego i ekonomicznego prognozowania wydobycia i użytkowania czystego węgla. Tom 1: Ekoefektywność technologii czystego spalania węgla. GIG, Katowice 2004.

Czaplicka-Kolarz K. i inni: Model ekologicznego i ekonomicznego prognozowania wydobycia i użytkowania czystego węgla. Tom 2: Ekoefektywność technologii czystego spalania węgla. GIG, Katowice 2004.

Bojarska-Kraus i inni: Ekoefektywność technologii. Praca statutowa GIG 2005, materiały niepublikowane.

LIST OF SELECTED U.S. PATENTS

[1] In-Situ Thermal Processing of a Coal Formation Using Pressure and/or Temperature Control. U.S. Pat. No. 6973967 (2005), http://www.freepatentsonline.com/6973967.html

[2] In-Situ Thermal Processing of a Coal Formation with a Selected Ratio of Heat Sources to Production Wells. U.S. Pat. No. 6959761 (2005), http://www.freepatentsonline.com/6959761.html

[3] Upgrading and Mining of Coal. U.S. Pat. No. 6969123 (2005), http://www.freepatentsonline.com/6969-123.html

[4] In-Situ Production of Synthesis Gas from a Coal Formation, The Synthesis Gas Having a Selected H_2 to CO Ratio. U.S. Pat. No. 6880635 (2005), http://www.freepatentsonline.com/6880635.html

[5] In-situ Thermal Processing of a Coal Formation to Produce Hydrocarbon Fluids and Synthesis Gas. U.S. Pat. No. 6761216 (2004), http://www.freepatentsonline.com/6761216.html

[6] In-Situ Thermal Processing of a Coal Formation Using a Controlled Heating Rate. U.S. Pat. No. 6749021 (2004), http://www.freepatentsonline.com/6749021.html

[7] In-Situ Thermal Processing of a Coal Formation to Form a Substantially Uniform, Relatively High Permeable Formation. U.S. Pat. No. 6742587 (2004), http://www.freepatentsonline.com/6742587.html

[8] In-Situ Thermal Processing of a Coal Formation to Pyrolysis a Selected Percentage of Hydrocarbon Material. U.S. Pat. No. 6712137 (2004), http://www.freepatentsonline.com/6712137.html

[9] In-Situ Thermal Processing of a Coal Formation with Carbon Dioxide Sequestration. U.S. Pat. No. 6763-886 (2004), http://www.freepatentsonline.com/6763886.html

[10] Process for Converting Gas to Liquids. U.S. Pat. No. 6,172,124 (2001), http://www.freepatentsonline.com/6172124.html

[11] Synthesis Gas Production System and Method. U.S. Pat. No. 6,085,512 (2000), http://freepatentsonline.com/6085512.html

International Mining Forum 2006, Sobczyk & Kicki (eds) © 2006 Taylor & Francis Group, London, ISBN 0415 401178

Underground Coal Mining in Russia
at the Beginning of the 21st Century and its Prospects

V.G. Gridin
"Siberian Business Council" Holding Company

V.I. Efimov
OOO "ФПГ Prokopevskurol"

V.A. Kharchenko
I.A. Stoyaniva
Department of "Economics of Nature Management" MSMU

ABSTRACT: Coal is a fuel, which is used by all the leading countries in the world to strengthen their national power systems. Moreover, the price for coal is usually significantly lower than those of gas and petroleum products. That is why it is economically more feasible to use coal rather than other types of fuel. It is not accidental that the share of coal in power generation in the USA, a country with the most strict market economy, is equal to 52%, in Germany, a country with a socially oriented market economy – 54%, in China – a country with a transition economy – 72%.

To a great extend the economic prosperity of Russia is determined by the status of the fuel/energy system of the country. Meanwhile, a high dependence of energy supply on the consumption of gas is a great problem. Increase in the share of natural gas in the fuel balance considerably reduces the reliability of the country's energy supply system and results in economically inefficient use of this non-renewable natural resource. It is therefore so important to increase the share of coal in the Russian energy production.

The share of coal in the total production of primary energy sources is 12.2% (Fig. 1). At the same time the production of oil in 2005 amounted to 470 196 thousand tonnes, gas – 640 633 mln m^3.

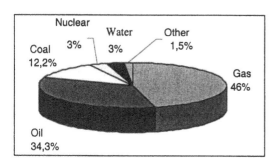

Figure 1. Share of coal in the production of primary energy sources

The total production of coal in Russia in 2005 amounted to 296 204 million tonnes. The dynamics of production are shown in Figure 2.

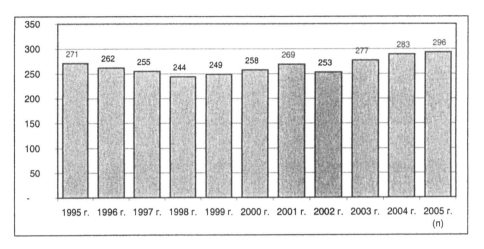

Figure 2. Production of coal in Russia, million tonnes

The main technical guidelines for the industry prescribed implementation of advanced methods for the development of more efficient production of coal by open-cast mining. In this context the share of underground coal production is now equal to 36%, which is 2 times less than it was in the former Soviet Union.

In 1994 the process of coal sector restructuring started. During 10 years the production of coal was closed at 202 particularly unprofitable enterprises with unsafe working conditions, including 187 underground mines and 15 opencast collieries. In the course of restructuring over 50% of the personnel working in the coal mining industry was retrenched, and the total number of discharged people exceeded 500 thousand. It should be mentioned that during those 10 years the government invested into this sector (including 2005) US$10.3 billion.

Coal production at the territory of Russia reached 427 million tonnes in 1988; by 1998 it dropped to 244 million tonnes. As from 1999 when the first stage of restructuring was completed the coal production started to grow. The major objectives of the Main Directions of Restructuring were basically achieved.

The coal industry operates currently according to the rules of the free market economy. Coal producing enterprises have been privatised. The share of private companies in the total volume of coal production is about 95%. Working efficiency in coal mining has reached its peak in the history of domestic coal industry – 138 t/month per one production worker, 100 t/month average for all mines. Labour productivity continues to increase. In the result of the restructuring the average daily production of coal from one face increased three fold. The total production capacity of operating underground and opencast mines amounts to approximately 350 million tonnes. At present the total production capacity of operating underground mines is equal to 140 million tonnes.

The perspectives of further build-up of high quality steam and coking coal production are mainly connected with the Kuznetsk Basin. In addition, some quantities of steam and coking coal are produced in the Pechorski basin and some types of hard coal in the Eastern Donetsk Basin.

Production of coal in the Kuznetsk Basin made up 57.4% of the total coal production in Russia, 48% of which is produced by the underground method.

So, bulk of the production comes from the Kuznetsk Basin.

The coal industry of Kuzbass makes a vast industrial-technological system including over 20 shareholding companies (corporations) and independent mines and opencast collieries. Current Kuzbass coal producers are represented by 60 mines and 36 collieries. Current Kuzbass coal producers are represented by 55 mines and 36 collieries.

The Kuznetsk Coal Basin is the largest coal-producing region in Russia, unique in the quality and the volume of its coal reserves. As a whole, the proven reserves of coal in the Kemerovo Region amount to approximately 91 billion tonnes (something like 46% of total reserves of Russia), the reserves of coking coal are equal to approximately 29 billion tonnes (about 73% of the total reserves of Russia). The coals of the Kuznetsk Basin are distinguished by their high quality: ash content – 8–22%, sulphur content – 0.3–1%, phosphorous – 0.002–0.1%, specific heat of combustion – 6000–8500 kcal/kg.

In 1997 the coal industry of the region reached its critical production level: it was hardly approaching 93.9 million tonnes, which indicates the decline in the production by almost 40%. As from 1998 coal production began to grow and in 2005 Kuzbass reached its maximum of 170 million tonnes, 81.6 million tonnes of which were produced by the underground method.

The most valuable types of coal are as a rule produced by the underground method. The main emphases are put on mechanised coalfaces using imported equipment. Nevertheless the performance of an average mining face is lower than in Germany or the USA, although in some cases it approaches generally accepted international level. One of the reasons of this situation is connected to complicated mining conditions. Besides, the mines mainly use worn-out obsolete equipment. Technical and technological parameters are lower than those achieved at German mines. For example, the coefficient of cutting machine productive time is 1,3–1,5 times smaller than in German mines, and the average longwall length is two times smaller. At the same time the installed power per employee is 2 times smaller, and the average length of an extraction block is 1,5–2 times smaller. Besides the mentioned factors there are other parameters that influence the operation of mines and production faces, such as unfavourable structure of mining facilities, lack of proper mining economy, ad-hoc nature of level development layouts, complicated ventilation systems, insufficient equipment reliability. So, the mode of the imported equipment operation results in prolonged emergency shutdowns and considerable reduction of the coefficient of productive time. Nevertheless, the average monthly output of a coal face equipped with "DBT" or "JOY" equipment is more than three times higher that that of traditionally-equipped longwalls. In stable operation conditions the longwall sets under consideration achieve over 100 thousand tonnes per face. Average monthly load on the complex-mechanized breakage face with local complexes makes 1481 tons/day, whereas average load on the longwall with "DBT" or "JOY" makes correspondingly 4894 tons and 5063 tons. And it should be taken into consideration that "DBT" or "JOY" complexes operate in more favourable mining-geological conditions and in the periods of stable operations the daily load on the breakage face is over 10 thousand tons.

US$1.052 million have been invested into development of the coal industry in the Kuznetsk Basin, and 3410 new work places were created. Wages increased by approximately 40% and reached at average US$526 per year. However, it should be mentioned that in some enterprises of the coal industry such as UK "Juzhkuzbassugol", "Raspadskaya" mine, enterprises of "Sibuglemet" Holding the workers of the main professional occupation (miners employed at production faces and developers) are paid US$1050–1400.

Almost US$2.105 million have been invested into the construction of new enterprises and installation of modern equipment at the operating coal entities, and 12 new mines and 16 opencast collieries have been set in operation since 1999.

The following four new modern mines were commissioned in 2005: "Vladimirskaya" "Novaya-2", unit 2 of "Anzherskaya-Juzhnaya" mine and "Kolmogorovskaya-2" and one open-cast colliery – Belorusskij. The total designed capacity of the new coal enterprises is 3.7 million tonnes, the total designed capacity of the underground mines is 3.2 million tonnes.

At present over 10 underground and opencast colliers with the total designed capacity of over 25 million tonnes are at various stages of design and construction. They will be put into operation before 2010.

According to the strategy of development for the period up to 2020, the yearly coal production is planned to reach 310–330 million tonnes by 2010, and 400–430 million tonnes by 2020.

The Government started to develop a detailed program of coal industry restructuring for the period 2006-2010. Its main goal is to create new highly productive coal enterprises, complete the technical works connected with closure of unprofitable mines, create safe working conditions as well as guarantee social support to the retrenched workers. As estimated about US$1.579 billion shall be required to realise the programme.

Apart from intensive development of the Erunakovski and Karakanski fields, the Kuznetsk Basin made a move towards the development of a new field – Tersinski that is in line with the strategy of economic development of Russia in the period up to 2020. This new field is rich in high quality coking coal, and its reserves are estimated to be in the range of 22 billion tonnes.

A stable tendency involving transition from the improvement of some industrial processes to the creation of mines that provide labour efficiency, safety and reduction of negative impact on the environment has been established according to the strategy of Russia relating to the coal industry (which is displayed in full in Kuzbass). The main technical directions required to realise the tasks are as follows:

1. Maximum concentration and intensification of mining operations especially longwalling by introducing automatic control system and implementation of processes without human involvement.
   Concentration of production envisages an output of over 3–5 thousand t/day per longwall with the transition to technological systems "longwall-mine" in the mines whose capacity is equal to 1.0–1.5 million tonnes. It is assumed that there shall be 6 mines of this type in the Kuznetsk Basin, 2 – in Pechorski and 3 – in Donetsk. The system "longwall-level" and "longwall-seam" shall be used for larger enterprises. Such capacities are available in Kuzbass (9), the Pechorski Basin (2) and Donetsk (2). In this case the average capacity of the mine shall be equal to 1.2 million t/year, the specific length of development workings shall be 56 m per 1000 tonnes of ROM coal.
2. Application of decisions relating to the preparation of mine field with a schematic set up that provides for complex degassing and effective ventilation.
   The panel method, predominantly with descending development of layers, shall remain to be used in cases of shallow dipping seams. The average longwall length should rise sharply (up to 300 m) irrespective of seam thickness and face run (over 1.5–2.0 km). Under the conditions prevailing in Russia extraction blocks of rectangular form of 250×2500 m in size shall make up 70–75% of the mineable seam area. The remaining parts of irregular form and smaller in size shall be mined-out by short (30–50 m) faces.
3. Using progressive technology and simplified mining operations with average output per longwall up to 5000 t/day.
4. Implementing industrial engineering.
5. Providing opportunity to use new technical solutions in the future.
6. Reduction of negative ecological impact of coal mines on the environment.
   The following shall be provided to solve this problem:
   – Preferential loans for pro-ecological programs.
   – Improvement of control and monitoring systems ensuring continuous and complete information pertaining to the state of the environment that allows to justify decisions, evaluate the consequences and develop recommendations concerning corrective measures.
   – Working out special development programs for each enterprise ensuring effective and safe nature management during the operation of the enterprise as well as after its closure.

Implementation of the above measures shall allow:

- To secure energy safety in the result of the increased coal production and its share in the balance of energy sources of Russia, thus reducing the dependence of the industry on natural gas.
- To increase the natural potential of Russia owing to the reduction of production and usage of non-recoverable primary natural resources by Fuel-Energy System of the country.
- To significantly improve technical and economic parameters of mines thanks to the introduction of the best experiences, concentration of mining, improvement of installed power per employee and the capacity of treatment equipment.
- To provide safety of mining operations through introduction of automated control systems and man-less production operations and preparation of mine fields ensuring complex degassing and ventilation.
- To improve social security of miners, to raise their wages thus making a miner's work attractive.
- To ensure ecological safety of coal producing regions by reconstructing old enterprises and constructing new ones that provide ecologically clean extraction and processing of coal, to make full use of the country's mineral wealth.

International Mining Forum 2006, Sobczyk & Kicki (eds) © 2006 Taylor & Francis Group, London, ISBN 0415 401178

Methods of Extraction of Thin and Rather Thin Coal Seams[*] in the Works of the Scientists of the Underground Mining Faculty (National Mining University)

Volodymyr Bondarenko, Roman Dychkovskiy
National Mining University, Dnipropetrovsk, Ukraine

ABSTRACT: The analyses of coal reserves in Ukraine are shown. The technological parameters of traditional mining are considered for given geological conditions. Features of mining technologies of augering, plowing, conveyor plowing, scraper plowing, rope saw extraction are considered. Technologies of the complex mechanized mining with cutter loaders in longwall faces are given.
   [*]Note: The Ukrainian coal seam thickness classification is as follows:
- < 0,7 m – very low coal,
- 0,7–1,2 m – low coal,
- 1,2–3,5 m – average coal,
- > 3,5 m – high coal.

KEYWORDS: Mechanized set, longwall, shortwall, plowing, conveyor plowing, scraper plowing, rope saw mining, augering, condition of rock, preparatory workings, parameters of mining system

INTRODUCTION

Maintaining the existing volumes of a coal production at the level of 80 million tons, and its increase up to 100 million tons requires mining very low and low coals. That part of the reserves is much greater than the average-width part of coal reserves in Ukraine. Therefore to improve the mining technologies of such reserves is the actual task to maintain the fuel balance of the country.

At the current production level the Ukrainian coal industry can effectively work for the next 200 years. Unfortunately, 65,9% of all the reserves prepared for mining are locked in seams whose thickness do not exceed 1,2 m (Fig. 1a). Prominent feature of the spatial arrangement of coal seams is that the main reserves are concentrated in flat seams (Fig. 1b). The world experience shows that mining of such reserves is not advisable because of high mining cost and low safety of miners in underground works. The approach in which the current price of coal forms the basis of nowadays researches is not correct. In the next 10 years the dynamics of fuel and energy balance in the world will change substantially. Coal's importance as an energy source raw material will grow essentially. This simply implies the necessity of research into the technology of mining of low and very low coals.

A lot of attention is given in scientific and technical and specialized literature to the application of complex mechanization in the mining of different thickness coal seams. Technologies of low and very low coal mining do not estimate enough the specific geological structure and tensions of the rock-mass. Also, no scientific parameters exist for the determination of principal technological methods of mining stress control.

In this work we try to generalize the potential for application of various mining technologies developed at the Faculty of Underground Mining (National Mining University, Dnipropetrovsk, Ukraine) to extracting such coal reserves and to estimate an opportunity of their application in various mining regions. Only traditional technologies of coal mining are presented in the article.

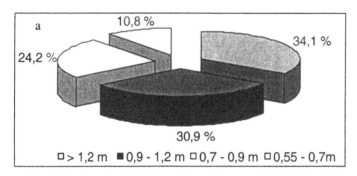

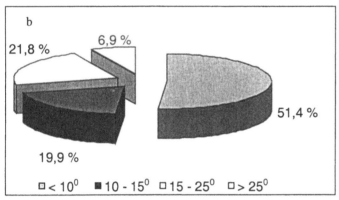

Figure 1. Distribution of coal reserves per thickness (a) and dip angle (b)

MINING TECHNOLOGIES OF PLOWING, CONVEYOR PLOWING, SCRAPER PLOWING AND ROPE SAW EXTRACTION

Domestic experience shows that these mining technologies are one of the most effective and safe for anthracite and coking coals and in outburst hazard conditions. They allow mining of average width, low and very low coals. The configuration of the mining equipment can be realized as a set or individual units. A plow coal cutting installation is situated between the face and the conveyor. It is rather different than in a mechanized longwall face equipped with a cutter loader.

The technological circuit of the plow mining method consists of a plow installation, scraper conveyor, jacks and mechanized or individual support. Such mining of coal is applied in coal seams characterized by relatively strong roofs allowing to be exposed over spans of 1,1–1,3 m for the period of duration of the extraction process.

At the beginning of the cycle the plow is situated in a niche and the hydraulic jacks press the conveyor to the face. The conveyor pans are used as an elastic beam and help the plow to constantly stay on the coal seam. Conveyor drive head are usually ahead of the line of the face by 0,2––0,5 m. It is necessary in order to maintain the best contact of the plow with the longwall face. Each hydraulic jack is installed between sections of the mechanized support and the frame of the conveyor. The plow moves between the longwall face and the conveyor, the extracted coal is loaded on the conveyor. Mining of coal is conducted in cuts of 150–200 mm. The coal mining is conducted from both sides of the face. Conveyor drive heads move after each pass of the plow. Outstripping niches 5–6 m in length and up to 3 m deep are arranged for their accommodation. The niches are prepared with the help of drilling and blasting or by jackhammers. Under the favourable geological conditions, the plowing drive heads are installed in the preparatory workings. Duration of mining is determined by the coal's strength properties and the plow capacity, and also the step of mechanized support movement.

Scraper plowing is simpler in technical and technological performance and is used in short-wall faces. Scraper plows are attached to cables and enable not only to extract but also to transport coal. Scraper plows and plows are intended to work only at constant face widths. In order to prevent face closure it is, as a rule, necessary to apply a backfilling of the goaf. Scraper plowing is characterized by low cost of the mining equipment. It is used with individual face supports. If scraper plowing installation is lost, it is necessary to prepare a new column.

In Ukraine, technologically advanced plows are industrially manufactured: KMC98M (plow mechanized set); UST4 and UST2M; and scraper plows: US2, US2M, US3 and other. Type of extracting tools on the unit is chosen according to the geological condition and strength of coal. Installations of tear and ram actions are most widely applied.

In conditions of inclined coal layers an effective way of mining is the application of the conveyor plowing. They are constructed in the form of a panelboard. The conveyor plowing is used in steeply inclined coal seams frontally to seam dip. The length of production face is 40–60 m.

Mining begins from a niche or a ledge and is conducted up towards the ventilation workings. Mining is done by working units equipped with picks, which move on a directing beam. The technological installation consist of a conveyor plow, a panelboard of mechanized support, hydraulic jacks, pump station and other devises.

Coal is transported along the face by scrapers fixed on round-link calibrated chain moving on the footwall of the seam. To the haulage way the mined coal is transported by gravitation. It is necessary to always leave some coal above the unloading chute to minimize the impact on the chute.

Rope saw coal extraction technology was developed by our scientists for steeply inclined coal seams. These technologies require creating pillars with lengths of 120 m and widths of 3–10 m. Along the contour of the pillar it is necessary to pass a steel traction rope 12–18 mm in diameter. Onto it cutting tools (saw) are attached. The tools can be: mills of diameter 40–90 mm, freely dressed on a rope; the mills connected by pieces of a chain, etc. Through deviating blocks traction ropes are connected to the drive of the rope saw.

The cutting tools of the installation are set in the bottom part of a coal strip at various distances from the foot of the seam. The rope with the saw is pulled by a hoist in a backward and forward motion (for 1–2 m, sometimes up to 4–5 m both ways). In this ways a narrow cut 30–90 mm wide is formed in the seam and this part of the coal face becomes weaker. Under the mining pressure and the weight the cut coal falls down on the footwall of the seam.

Coal is removed from the face and into a haulage way below by gravitation. Then it is transported according to the mine transport system.

The face is not supported. To avoid roof caving the room sizes are designed for different length and width. Mining is conducted very quickly.

Ventilation of the working space is unstable. Coal extraction is conducted in a neutral environment.

Mining technologies of plowing, scraper plowing, conveyor plowing and rope saw extraction have limited application. Strength, water inflow, gas-bearing of coal and stability of surrounding rock are the main factors which confine their usage. Therefore, our scientists developed the technical and technological conditions for traditional and novel technologies of mining. Technological alterations to the technology of mechanized mining are made too.

TECHNOLOGIES OF MECHANIZED MINING SETS
WITH CUTTER LOADERS IN LONGWALL FACES

The term "longwall mining" applies to the situations when the length of a face is more than 10 times longer than the width of the working space. Such technologies have wide application in the countries of Europe, Asia and are partly used in the mines of North America. They are used in a wide range of the coal thickness from 0,9 to 5,0 m and at different angles of dip. Mining is done by cutting along the longwall face.

This technology of mining is characterized by: constancy of working space, satiation by mining technological processes mechanization equipment, continuous ventilation, the observance of safety standards on the content of harmful gases and dust particles in the mine atmosphere. In most technological systems cyclic organization of works is used in coal mining.

This mining technology is well described in specialized and technical literature. Therefore we will not depict the technological circuit of the mining processes at a longwall face. In low and very low coals the technological processes are carried out with the permanent and partial presence of people in the working space of the face.

In the second case, technological processes are handled by the systems of automatic management and control from the preparatory workings. The people in the face are present only when attending to emergency situations. Negative side of such mining technology is absence of direct contact of man in the system „mining worker – technical equipment – geological environment".

In accordance with the requirements of the Ukrainian regulations, the largest possible thickness of coal seam suitable for mechanized mining with the permanent presence of people in the face is 0,9 m. Mining of thinner seams means extracting roof or footwall rock.

For quality mining of coal technologies of separate and selective coal and waste rock mining were developed. The technological solutions enable to separate waste rocks and coal at the face. Modern mechanized sets are utilized in this mining technique. The sequence of the structural elements of this technology in coal and waste rock mining and the duration of technological cycle can be separated in time. The rock also can be extracted simultaneously with coal. For this purpose it is suggested to use scraper conveyors with two working flights: upper – for the rock, lower – for the coal.

The technology of longwall coal mining is of a flexible structure. The number of elements composing the structure can change depending on geological conditions and the accepted elements of mining mechanization. At any case, such technology consists of: coal extracting, waste rock extracting (together with coal to another conveyor flight or after coal extraction), loading of coal and rock, transport of coal and rock, support of the working space, movement of scraper conveyor, processes at the ends of the longwall face, management of the mining stress.

Implementation of the mining is carried out in connection with the auxiliary operations, organization of work, providing the face ventilation and absolute observance of safety rules.

Negative features of the selective and separate mining of coal and rock are large power demand from coal and rock mining and transport. For these technologies it is possible to apply only narrow cutter loaders. All this results in production cost of coal rapidly rising. Also the questions of goaf backfill and rock storing on the surface are not solved.

AUGER COAL MINING

This mining technology is conducted with the use of special augering installations from preparatory workings. They allow mining coal from pillars, geologically disturbed areas, zones of high mining stress and other uneconomic coal reserves. The last modification of the augering method can be applied in gas-bearing seams (up to 20 m^3/t.d.m.) and in condition of the surrounding rocks with below average strength. Ventilation of the mining site is conducted in common manner. Direction of movement of drill-rod along the seam is carried out with the use of special sensors, which automatically correct the coal mining.

Domestic machine-building industry of mining equipment developed a number of augering machines. They are intended for coal drilling from gallies up to 85 m long and 2 m wide in seams from 0,55 to 1,2 m wide without the presence of auxiliary personnel in the mining space and without support of the goaf. The system is used to mine coal in coal seams where fire, gas and coal dust explosion hazards are not present and for coal with resistance of up to 350 KN/m.

Drilling of mining stripes is done from the preparatory workings with cross-section of no less than 11,2 m^2, developed with footwall ripping of not less than 0,6 m. The limits of coal seam dip angle are as follows: up – max. 25°; down – max. 10°. The tilt of preparatory workings must be less than 3°.

The setup consists of an augering machine, auger drill, and system of ventilation. Auger drill of the augering machine is equipped with three boring bits. The rotation of auger drill is carried out by two separate hydraulically operated drives on the frame. Special skids are used to move the equipment and place it in the working position. Hydraulic jacks ensure the steady disposition of the augering in the preparatory workings. For the mechanization of docking and undocking of the auger drill the installation has a centring block, which fixes the drills in the direction of the stripe. A monorail transport system is used to help with the assembly of the augering elements.

The removal of the drilled coal is done by a scraper conveyor located in the preparatory workings and further according the technological transport system of the mine.

Power installation of the set includes pump station, which is used for augering hydraulic system; blocks of management apparatus on the frame operators panel; electric station (type SUV-350AV or other), which feeds all the units using electricity of the set.

The augering moves along the preparatory workings by wire pulling. The step of movement between stripes is 2,6–3,1 m depending on the geological conditions.

The ventilation of the working area is provided by ventilation pipes. Necessary quantity of air in the vent pipeline is controlled by the apparatus of gas protection (ATZ-1 or other). The system of ventilation consists of a fan and a pipeline, which is extended accordingly, with the progress of mining.

Application of augering has limited area of application. Presence of stable rock conditions is the obligatory condition of normal work of augering. At presence of weak rocks it is not possible to retain the auger drill within the limits of the coal seam and the sensors controlling the position of the augering machine along the coal seam become not capable.

CONCLUSION

Application of highly productive mechanized sets, different mining cutter loaders, plows, scraper plowing mining and other mechanization facilities with parameters corresponding to the mining and geological conditions of coal are the main directions in the development of coal mining technology. Mine practice shows that the variety of actual geological situation requires adequate change in technical equipping and effective management of mining stress and control of mine ventilation. The offered technical and technological solutions allow to intensify the process of coal mining and to obtain coal at rational production cost.

International Mining Forum 2006, Sobczyk & Kicki (eds) © 2006 Taylor & Francis Group, London, ISBN 0415 401178

Minimizing Coal Losses When Extracting Thin Coal Seams with the Use of Auger Mining Technologies

Iryna Kovalevska
National Mining University. Dnipropetrovsk, Ukraine

ABSTRACT: The paper presents the mining experience obtained in working in twice undermined coal seam by auger technology. The parameters of the suggested technology are determined by the dividing coal pillars according to the functional assignment. The results of scientific and practical work done at "Dobropolska" mine are given.

KEYWORDS: Auger mining, losses of coal, basic & crosshole pillars, arch of natural balance

The negative phenomena of auger mining are high losses of coal and the absence of experience in mining of underworked thin layers. By the decision of the Ukrainian Coal Mining Ministry and in co-operation with the Donetsk Coal Mining Institute (DonUGI) a research-and-production project for auger mining technology of twice under-mined seam l_4 has been created at "Dobropolska" mine. Seam l_4 was mined in rather difficult geological conditions. Three groups of industrial experiments have been carried out in this seam on different sites. The results of the research and measurements justified the choice of auger mining for these conditions. It also showed the necessity to improve the parameters of this technology by dividing functional, basic and crosshole coal pillars. The last one is left with a variable width along the length of coal mining.

During undermining of seam l_4 a huge zone of full rock movement was created. Its area is defined by the size of the extraction works in the seams situated below and the height, approximately, is equal to 0,8 of the area of the mining works in the underlying seams [Zborshchik *et al.* 1991]. The rocks contained in this zone are destructed and divided into separate pieces and blocks of various sizes (small enough in comparison with the sizes of this zone). The character and properties of this condition are close to those of loose material (or to inconsistent rocks). It has essentially complicated the mining of seam l_4 not only by traditional technology with the application of cutter loaders but also for the auger technology with coal pillars with stable width.

The newly offered auger technological system is shown on Figure 1. Crosshole pillars 1 of small width l_1 are left between drilling holes during coal mining. They perform two basic functions. First, during drilling of the next hole the neighboring crosshole pillar takes the load from the weight of rock inside the arch of natural balance and by that protects the drilling mechanism. Second, the crosshole pillar is considered as a mining stress control facility: during further drilling the arch of natural balance increases above the pillar (load from the weight of rock increases) and the pillar is destroyed by the weight of the rock.

The work of a crosshole pillar during crushing is accompanied by enough yields to promote the reduction in mining stress on the part of the seam under extraction. All loading from the weight of rock in the arch of natural balance is transferred to the footwall. Reduction of the mining stress in the part of the seam under extraction promotes protection of the drilling mechanism from clamping. After drilling of certain number of holes along the length l_3 of the working a basic pillar 2 with the width l_2 is formed. This width prevents it from clamping during all the period of column mining (before cutting the following basic pillar on border with coal seam). The main function of the basic pillar is in recognition of roof weight, formation of the arch of natural balance between the basic pillars, reduction of the load on crosshole pillars and support of preparatory workings.

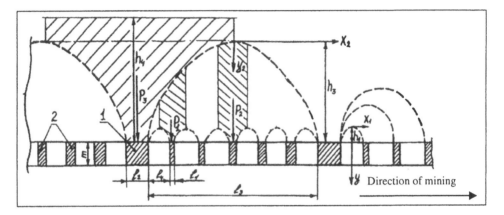

Figure 1. Technological and settlement circuit on cutting
and parameters calculation of basic 1 and crosshole 2 coal pillars

Due to the arch of natural balance loading between basic coals pillars, load P on crosshole pillars depends on the distance from a basic pillar. Minimum load P_1 on crosshole pillar will be in situation, when it is nearest to a basic pillar. The maximum load P_2 is in the center of the arch of natural balance (Fig. 1). Loads on other crosshole pillars have intermediate values.

Loads P_1 and P_2 can be determined in the following way. A column of the destroyed rock operates on a crosshole pillar. This column is limited at the bottom by the surface of the arch of natural balance between neighboring crosshole pillars; at the top – by the surface of the arch of natural balance between two neighboring basic pillars. This is recommended to accept the width of the column of the loading rock at the center axes of l_4. Thus, the width of a column of rock will be equal to $l_4 + l_1$. Research [Zborshchik 1978] proved experimentally that the arch of natural balance in the broken rockmass in roof of preparatory workings is quite well described by the theory of prof. M.M. Protodjakonov. Such conclusion is quite logical as this theory was developed for inconsistent (loose) rocks. Actually, the rocks in a roof of twice undermined seam are in such condition. There is enough evidence that the use in these geological conditions the M.M. Protodjakonov theory (theory of natural arch creation) defining the sizes of natural balance between both cross-hole and basic pillars are well grounded.

According to this theory the arch of natural balance is described by the equation of a parabola:
– for distance l_4:
$$y_1 = \frac{2x_1^2}{l_4 f},$$
– for distance l_3:
$$y_2 = \frac{2x_2^2}{l_3 f},$$
where f – factor of compaction of rock in the roof of the seam.

For loose roof rocks this factor is defined only by the angle φ of internal friction ($f = tg\varphi$). So, the equations of a vault surface of natural balance above distances l_4 and l_3 will be:

$$
\left.
\begin{aligned}
y_1 &= \frac{2x_1^2}{l_4 tg\varphi}; \\
y_2 &= \frac{2x_2^2}{l_3 tg\varphi}
\end{aligned}
\right\}
\tag{1}
$$

According to (1), we can calculate load P_1 on first crosshole pillar:

$$
P_1 = \frac{\gamma}{2l_3 tg\varphi}\left[(l_4 + l_1)(l_3 - 4l_4 - l_1)^2 - \frac{l_4^3}{3}\right]
\tag{2}
$$

where γ – average volumetric weight of rocks inside the arch of natural balance.

The maximum load P_2 on crosshole pillar is defined similarly. When it is situated in the middle of distance l_3, in account that $h_3 = \dfrac{l_3}{2tg\varphi}$, it will be:

$$
P_2 = \frac{\gamma}{2tg\varphi}\left[l_3(l_1 + l_4) - \frac{l_4^2}{3}\right]
\tag{3}
$$

Formulae (2) and (3) are the basis for further choice of pillar widths.
Load P_3 on the basic pillar is defined by the formula:

$$
P_3 = \frac{\gamma l_3}{2tg\varphi}\left(l_2 + \frac{l_3}{3}\right)
\tag{4}
$$

Obvious direction towards the reduction in losses of coal in pillars is the reduction of their width (l_1 and l_2) till the minimum permissible sizes. As to crosshole pillars parameters I proposed the following effective solution. The pillar nearest to drilling hole should keep stability during the whole period of the drilling works. Apart from that it is in a relaxing zone of the arch of natural balance of the seam roof. As soon as the production face will depart from pillar by two – three drilling holes, the necessity to keep it steady disappears. Thus, crosshole pillar carries out the functions of protection for drilling hole from jamming. Here is the zone of smallest load.

As to the basic pillar, we can get such distance l_3 between basic pillars, when losses of coal V will be the smallest. Taking into account that coal losses in pillars depend on widths l_1 and l_2, which are calculated from geotechnical characteristics, we can get the function of the purpose:

$$
\frac{V}{m(l_2 + l_3)}(l_5, l_6, r, \varphi, \gamma, l_4, m, D, \sigma) \to min
\tag{5}
$$

where D – agency diameter of an auger installation, σ – coal strength, r – radius of the preparatory working, l_5 and l_6 – the length of a drill hole and the length of a basic pillar.

Expressing width of the pillar through symbols l_1 and l_2, it is possible to simplify the function of the purpose (5):

$$
\frac{nV_1 + V_2 + V_3(n + 1)}{(l_2 + l_3)m} \to min
\tag{6}
$$

where V_1, V_2 and V_3 – losses of coal in crosshole pillars, basic pillars, and in the drill hole (usually a pack of coal is left in the roof of the seam), n – number of crosshole pillars.

For definition of parameters V_1, V_2, V_3 and n the following expressions are received:

$$\left.\begin{array}{l} V_1 = l_1(2m-D)+mD-\dfrac{\pi D^2}{4}; \\[3mm] V_2 = l_2(2m-D)+mD-\dfrac{\pi D^2}{4}; \\[3mm] V_3 = 2Dm-\dfrac{\pi}{2}D^2; \\[3mm] n = \dfrac{l_3-3D}{l_1+3D} \end{array}\right\} \qquad (7)$$

Solving in common the equations (7) and (6), we can come to the criterion of the minimum coal loss:

$$\frac{1}{l_2+l_3}\left\{\frac{l_3-3D}{l_1+3D}\left[l_1\left(2-\frac{D}{m}\right)+3D-\frac{3}{4}\pi\frac{D^2}{m}\right]+l_2\left(2-\frac{D}{m}\right)+3D-\frac{3}{4}\pi\frac{D^2}{m}\right\} \to \min \qquad (8)$$

To determine the rational value l_3 of the distances between basic pillars according to the formula (8), it is necessary to define widths l_1 and l_2, and then to search (8) for the minimum. Therefore the following steps are the calculation of parameters of pillars according to their functional purpose.

Among the existing methods of such calculations there are two basic ways: according to the admitted stress equal to the strength on compression $\sigma_{cж}$ of the pillar material; and according to the admitted load $P_{нес}$, which is determined from the stress condition of the pillar.

In the first way the estimation of pillar stability is done on the basis of comparison of some average (in square of pillar) tensions σ with durability of the pillar material in compression $\sigma_{cж}$ along one axis. This parameter is determined from laboratory samples of the pillar material. Such methodology is chosen because defining mechanical properties of pillars directly in a mine is rather inconvenient. It is necessary to take into account the scale factor, which mentioned various geological and technological discontinuities. It is necessary to enter correction factors into the results of laboratory tests. The most important of them is the form factor K_ϕ. It shows the relationship between the strength of a pillar in a cubic and other than cubic (prismatic) form. For the crosshole pillar with the proportion of pillar height, equal to the thickness m of the seam, to its width l_1 in the most typical range:

$$1 \le \frac{m}{l_1} \le 3,$$

it is possible with sufficient accuracy to recommend Erofeeva's formula:

$$K_\phi = 1,15 - 0,15\frac{m}{l_1} \qquad (9)$$

The specified range allows estimating pillar parameters from the first method, i.e. to estimate its stability from the strength of coal in uniaxial compression $\sigma_{cж.y}$. It is necessary to take into account that undermining breaks coal seam, that's why it is necessary to do the calculations from residual uniaxial compression strength of coal $\sigma^0_{cж.y}$. Its value is determined in laboratory conditions.

Basic pillars, bearing the load from most part of rock in the arch of natural balance, have much more width l_2, which frequently exceeds pillar height. So, for basic pillars the ratio $m/l_2 < 1$ is fair. A complex stress condition exists in the pillar [Ruppenejt et al. 1960] so it is inadmissible to estimate it from the first method. Stability of the pillars is expediently assessed by the second method. Actual load on a pillar is compared with its bearing capacity $P_{нес}$. Bearing capacity of a pillar is defined on the basis of the theory of limited balance. It means that pillar material passes into a condition of limited balance [3].

Thus pillar-bearing capacity is determined (in our case basic) in view of rectilinear bending around the Mohre circles:

$$P_{\text{нес}} = 2cl_2 \left[2 + F\left(\frac{l_2}{m}, \varphi_y \right) \right],$$

where c and φ_y – cohesion and angle of internal friction of coal (cohesion is defined in view of the partial crushing of the seam), $F(l_2/m, \varphi_y)$ – the tabulated function, reflecting a high stress condition in pillar; it is defined in [3].

According to the above-stated, it is possible to write down two conditions for the choice of cross-hole pillar width l_1. They determine the minimum value needed for pillar stability l_{1min} near drilling hole; and the maximum value of pillar width l_{1max}, when it is in the middle of the distance l_3.

Then it is possible to write down a system of inequalities:

$$\left. \begin{array}{l} P_1 \leq K_\phi \sigma^o_{\text{сж.y}} l_{1min} ; \\ P_2 > K_\phi \sigma^o_{\text{сж.y}} l_{1max} \end{array} \right\} \tag{10}$$

The first inequality (10) of the system is the determining. Having solved it together with the formula (2), we will get

$$l_{1min} \geq \frac{\gamma}{2K_\phi l_3 \sigma^o_{\text{сж.y}} \operatorname{tg}\varphi} \left[(l_4 + l_{1min})(l_3 - 4l_4 - l_{1min})^2 - \frac{l_4^3}{3} \right] \tag{11}$$

The second inequality is for verifying, whether finding out the minimum pillar width l_{1min} exceeds such value, in which crosshole pillar remains stable being in the middle of the distance l_3. It is practically impossible to obtain from the analysis of the settlement circuit such variant (when $l_{1max} < l_{1min}$), as load P_1 will always be smaller than load P_2 on the crosshole pillar in the middle. Nevertheless it is interesting to see what safety factor can be chosen near a crosshole pillar so that in the middle of the distance l_3 all the same occurred it pressing. For this purpose we shall determine the maximum pillar width l_{1max} by solving the second inequality of system (10) and expression (3):

$$l_{1max} < \frac{\gamma l_4 \left(l_3 - \dfrac{l_4}{3} \right)}{2\operatorname{tg}\varphi K_\phi \sigma^o_{\text{сж.y}} \left(2\operatorname{tg}\varphi K_\phi \sigma^o_{\text{сж.y}} + \gamma l_3 \right)} \tag{12}$$

Both equations (11) and (12) are transcendental and their solutions are found only approximately by numerical methods. For convenience, the calculation of parameters l_{1min} and l_{1max} is conducted on the basis of the numerical analysis of the equations (11) and (12). Corresponding monograms were created [Bondarenko et al. 1999].

Smallest allowed width of a basic pillar l_2 in condition of its stability is determined by the equation (13):

$$l_2 \geq \frac{\gamma l_3 \left(h_4 - \dfrac{l_3}{3\operatorname{tg}\varphi} \right)}{2c \left[2 + F\left(\dfrac{l_2}{m}, \varphi_y \right) \right] - \dfrac{\gamma l_3}{2\operatorname{tg}\varphi}} \tag{13}$$

where h_4 – height of the column of undermined rock above the basic pillar.

However equation (13) is transcendental, in an obvious kind is not solved, and width l_2 we can determined by monogram [4].

On Figure 2 the schedule of dependences of smallest allowed width of crosshole pillar l_{1min} on the distance between basic pillars l_3 is given for different mechanical characteristics of rock.

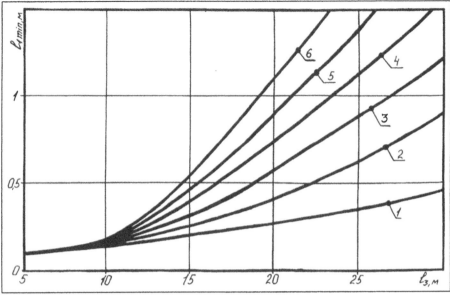

$1 - \sigma^{\circ}_{\text{сж.y}}\,\text{tg}\varphi = 6\,\text{МПа}$; $2 - \sigma^{\circ}_{\text{сж.y}}\,\text{tg}\varphi = 5\,\text{МПа}$; $3 - \sigma^{\circ}_{\text{сж.y}}\,\text{tg}\varphi = 4\,\text{МПа}$;

$4 - \sigma^{\circ}_{\text{сж.y}}\,\text{tg}\varphi = 3\,\text{МПа}$; $5 - \sigma^{\circ}_{\text{сж.y}}\,\text{tg}\varphi = 2\,\text{МПа}$; $6 - \sigma^{\circ}_{\text{сж.y}}\,\text{tg}\varphi = 1\,\text{МПа}$

Figure 2. Dependence of minimum width $l_{1\text{min}}$ of crosshole pillar from distance between basic pillars l_3

Pillar width $l_{1\text{min}}$ changes depending on the distance between basic pillars l_3 and for $l_3 \leq 20$ m it changes from 0,55 m to 1,1 m at parameter $\sigma^{\circ}_{\text{сж.y}}\,\text{tg}\varphi \leq 4$ MPa (Fig. 2). An increase in the distance between basic pillars l_3 leads to essential losses of coal in crosshole pillars. From this point of view it is expedient to suppose that the distance between basic pillars l_3 will be 18–20 m (that corresponds to drilling 9–10 holes between them). The analysis of dependence $l_{1\text{min}}$ from l_3 confirms the expediency of cutting crosshole pillars of variable width as a multiplication of coal drilling between neighboring basic pillars. It allows reducing losses of coal together with the preservation of the functions stipulated by the technology.

On Figure 3 dependence of smallest allowed width of basic pillar l_2 on the distance between them l_3 is shown in condition of variable seam thickness m with other parameters fixed. Connection l_2 with l_3 has non-linear character. So, at $5\ \text{m} \leq l_3 \leq 15$ m there is a substantial growth of the required width of the basic pillar l_2 (from 0,95 to 1,3 m at $l_3 = 5$ m to 1,3–1,85 m at $l_3 = 15$ m at seam thickness variation from 0,7 m to 1,2 m). At further increase of the distance l_3 ($l_3 > 15$ m) the curves occur more flat l_2 (l_3) and increase of width of basic pillar l_2 occurs less intensely, and after $l_3 = 20$ m – within the limits of several percent.

The described law says that with an increase in width l_2 of basic pillar its bearing capacity sharply grows. Intensity of width growth l_2 at increase in distance l_3 depends on mechanical properties of coal and roof rocks.

Thus, it is established, that rational value of smallest allowed width of crosshole pillars is changeable l_3 between basic pillars. Crosshole pillars of variable widths allow reducing losses of coal at auger mining. Also, the dependence of losses of coal from distance l_3 between basic pillars was revealed. Thus the presence of minimum coal losses was established at certain values l_3. Depending on geometrical parameters of auger technology and mechanical characteristics of coal and roof rocks rational value of the distance between basic pillars changes within limits from 10 to 20 m.

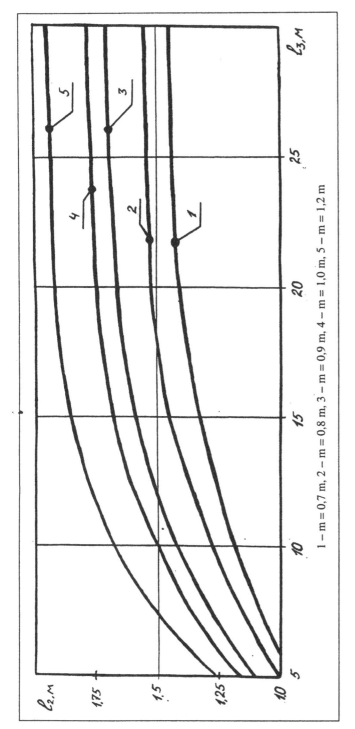

Figure 3. Dependence of smallest allowed width l_2 basic pillar from distance between them

1 – m = 0,7 m, 2 – m = 0,8 m, 3 – m = 0,9 m, 4 – m = 1,0 m, 5 – m = 1,2 m

REFERENCES

Zborshchik M.P., Nazimko V.V.: Deep Mines Workings Protection in Zones of Unloading. KYIV, Technicka 1991, p. 247.

Zborshchik M.P.: Deep Mines Workings Protection in Goaf. KYIV, Technicka 1978, p. 176.

Ruppenejt K.V., Liberman Yu.M.: Introduction in Mechanics of Rocks. M.: Gosgortehizdat 1960, p. 356.

Bondarenko V.I., Kovalevska I.A., Bugajev I.V., Simanovich G.A.: Reduction in Losses of Coal at Augering Mining of Thin Coal Seams. The Scientific Bulletin of National Mining Academy of Ukraine 1999, № 4, p. 36–39

International Mining Forum 2006, Sobczyk & Kicki (eds) © 2006 Taylor & Francis Group, London, ISBN 0415 401178

Geomechanics: History, Modern State and Prospects of Development

Olexandr Shashenko
National Mining University. Dnipropetrovsk, Ukraine

Tadeusz Majcherczyk
AGH – University of Science and Technology. Cracow, Poland

ABSTRACT: The history and the main steps of development of geomechanics is considered. The modern state of modelling of the mechanical processes taking place in rock mass is given.

KEYWORDS: Geomechanics, mathematical modeling, mechanical processes, longwall method, rope saw mining, gas-dynamics manifestation

The history of development of geomechanics covers a relatively short period of time from the beginning of the last century till our days. From 1904 M.M. Protodiakonov publishes a series of articles dedicated to the problems of mining stress, which later become the base of his monograph "Mining rock stress to mining support" published in 1907. That was the first great theoretical work in the field of rock mechanics. It considered the hypothesis of natural balanced arch over mining works and described the calculation method of supports for shallow mines. So the development of geomechanics in Russia began.

W. Budryk's works in the field of mathematical modelling of mechanical processes taking place in the rock mass during the extraction of layers of coal with the longwall method became the beginning of geomechanics development in Poland. The ideas of W. Budryk were later advanced in works of his disciples: A. Salustowicz, S. Knothe, J. Litwiniszyn, etc.

In 1911 T. Karmman's works on research of mining rocks in conditions of three-dimensional compression appeared. He also checked the correctness of the Moor's theory.

In the years 1920–1950 O.F. Graf, F.A. Belaenko, A.A. Borisov, W. Budryk, F.K.T. Van-Iterson, R. Kvapil, M.I. Koyfman, G.N. Kuznecov, V.D. Slesarev, A. Salustowicz, D.S. Rostovcev, A.I. Celigorov, P.M. Cimbarevich and other authors published their works, mostly dedicated to the problems of mining stress and its influence on support, investigation of physical and mechanical rock properties, rockmass stability. Research projects of this period are mostly based on simple models of mediums and explored objects.

Later, in the years 1950–1980, fundamental works in the field of mining stress based on limiting condition theories (V.V. Sokolovsky, N.N. Maslov, G.L. Fisenko, V.N. Zemisev, etc.), on theories of elasticity, plasticity and creep (F.A. Belaenko, V.V. Vinogradov, J.S. Erjanov, L.V. Ershov, B.A. Kartozia, A. Labass, U.V. Liberman, G.G. Litvinsky, L.Y. Parchevsky, A.G. Protosenya, K.V. Ruppeneyt, M.M. Rozovsky, R. Fenner, A.N. Shashenko, E.I. Shemyakin and others) were published. It was in this period when papers, in which the natural rock mass was considered on the basis of statistic models (I.V. Baklashov, M.A. Dolgih, V.V. Matvienko, L.Y. Parchevsky, K.V. Ruppeneyt, A.N. Shashenko, V.I. Sheynin and others) appeared.

In the same time deep investigations in the field of rock destruction, including controlled loading mode, were conducted in such well-known geomechanic scientific research centres as VNIMI, Mining Institute named after Skochinsky, IGTM NAS of Ukraine, Dnipropetrovsk Mining Institute, Doneck Polytechnic Institute, Leningrad Mining Institute, Moscow Mining Institute and others leading scientific research centres in SAR, Japan, England and Poland.

Reviews of these researches were done by V.V. Vinogradov, Z. Bieniawski, A. Cook, A.N. Stavrogin, L.I. Baron, E.I. Ilnitskaja, J.M. Kartashov, Z. Kleczek, G.N. Kuznetsov, B.V. Matveev, T. Majcherczyk, A.G. Protasenja, V.V. Rzhevsky, V.S. Jamshchikov, G.T. Kirinichansky and other authors. They allowed creating new, qualitative rockmass models, putting forward and solving elastic-plastic tasks in the new statement.

By that time an independent line of rock mechanics was formed. It was connected to the research into the gas and dynamic manifestations of the mining stress: sudden burst of coal, rock, gas, seismicity (S.G. Avershin, B. Drzezla, Z. Kleczek, J. Dubinski, M. Chudek, L.N. Bikov, A.N. Zorin, I.M. Petuhov, V.V. Hodot, E.I. Shemyakin and others).

Thorough studies of surface movement under the influence of mining were carried out, which are represented by the works of, for instance, S.G. Avershin, V.F. Galahov, V.N. Zemisev, M.A. Iofis, S. Knothe, J. Litwiniszyn, R.A. Muller, V.I. Mjakenkij and other authors.

In that period, such frequent in mining workings a phenomenon as floor heave was explored (P.M. Cimbarevich, V.D. Slesarev, A.P. Maximov, Y.Z. Zaslavskiy, I.L. Chernyak, V.A. Litkin, V.T. Glushko, A.N. Shashenko and others).

A large piece of work on studying mining stress appearing in natural conditions was carried out by scientific collectives under the supervision of V.T. Glushko, Y.Z. Zaslavsky, A.N. Zorin, A.G. Krupennikov, K.V. Koshelev, A.P. Maximov, T. Majcherczyk, I.D. Nasonov, G.I. Pokrovskij, I.L. Chernyak and others.

In the same time, wide laboratory research on modelling geomechanical processes (G.N. Kuznecov, G.I. Pokrovskij, V.F. Trumbachev and others) was undertaken.

In the 70's and 80's, determinative models in geomechanics were explored to a satisfactory degree. As a rule, they were carried out in order to analyse the stability of a single long gallery, which was the main underground construction structure unit. The determinative model required idealization of the research object (mining working, rock mass), and boundary conditions. This circumstance allowed using solid mechanics techniques and obtaining distinct solutions of the problems. But comparing actual measurements with analytical calculations showed that high idealization of the object in geomechanics led to such a situation: a range of results in actual conditions fluctuated around the forecasted result, corresponding to a single distinct result of a theoretical calculation. The range of factors that were not included into the physical model due to their unimportance (according to the initial viewpoint) was the main reason for these differences.

This circumstance caused a distrust among the practising mine designers and engineers, compelling them to correct analytical calculations and to accept technological and technical parameters intuitively, first of all basing them on the experience gained in similar underground mining conditions. This approach very rarely results in making optimal engineering decisions, which leads to the increase of expenditure on maintaining the mining workings or superfluous support stability and higher costs of support. It is especially seen in the mines, which were built in difficult mining-geological conditions in the Eastern and Western Donbass, Krasnoarmeysk coal region, Lviv-Volyn coal region etc.

Obviously it is the unconformity of determinative models with real objects that is the reason for such a situation. It is impossible to forecast behaviour of such difficult objects like rock mass containing a mining excavation using only models where everything is defined and there is no place for variability. Any highly organized system is formed and exists effectively when it has some internal instability. Then the strict connection of one structural unit, existing by action of some common physical laws, would not make any strong restrictions to other elements, which should have an opportunity to be changed in a random way.

This shows that geomechanical research should be based on stochastic models, which are more complicated and reflect fluctuation of parameters around determinative values. Understanding this problem led to the occurrence of some research projects based on probability models during the last 15–20 years (K.V. Rupeneyt, L.Y. Parchevsky, A.N. Shashenko, S.B. Tulub, V.P. Pustovoytenko, G.T. Rubec, E.A. Sdvijkova and others). The spectrum of these works is considerable: strength-strain state of rock, floor heave, expense optimisation for long-life underground constructions, deformation of support, quick analysis of management decisions etc.

Now we can surely say that statistical geomechanics became an independent branch of science dealing with the mechanical processes taking place in rocks.

At the end of the 80's geomechanical models became so difficult that using analytical methods (as they are) ceased to be effective. At that time personal computers became more powerful and accessible for carrying out research projects. With the development of personal computers numerical methods became very effective and popular, as it may be judged from the research projects by B.Z. Amusin, J.S. Erjanov, T.D. Karimbaev, S. Krauch, L.V. Novikova, E.A. Sdvijkova, A. Starfield, B.A. Fadeev and others.

This direction in the development of geomechanics is considerable and perspective.

The combination of continuum and discrete medium methods with new geoinformation technologies, geophysics, theory of mathematical statistics constitutes the basis for creating a system of global geomechanical computer monitoring. Then the exploration of underground space can become more effective and safer.

International Mining Forum 2006, Sobczyk & Kicki (eds) © 2006 Taylor & Francis Group, London, ISBN 0415 401178

Speed of Roof Rock Separation and a Type of Working's Support

Tadeusz Majcherczyk, Piotr Małkowski, Zbigniew Niedbalski
AGH – University of Science and Technology, Cracow. Poland

ABSTRACT: The paper presents measurement results of speed of roof rock separation for three selected headings protected with various types of support, i.e. individual roof bolting, combined system of prop support and roof bolting, and prop support. The measurements were carried out for a period of dozen to several dozens of months. Separations were recorded with the use of low and high multilevel telltales as well as extensometers. Speed of separations was determined in relation to the time of measurement. Recording separations for a long period of time allowed for an estimation of stability of the analysed workings as well as for relating movement of roof layers to various types of support applied in their construction.

1. INTRODUCTION

Speed of roof rock separation is one of the basic factors affecting stability of headings and it is most often understood as a quotient of separation value and an assumed unit of time [Butra & Orzepowski 2001]. The measurement of separation speed can be carried out with the use of special measuring instruments or standard telltales and extensometers if the changes around the heading are going on for a long period of time [Butra 1998, Fabianczyk 2001, Majcherczyk *et al.* 2005]. A considerable intensiveness of the phenomenon requires the use of a proper method of limiting roof rock separation, thanks to which it is possible to preserve the working's stability [Gale & Blackwood 1987]. Apart from geological conditions and mining situation, a type of support applied in the working is one of the crucial factors affecting changes of separation in time [Majcherczyk & Niedbalski 2002a, 2002b, Szymiczek *et al.* 2001, Fabich *et al.* 2000]. Separation speed will not influence the stability of heading with a prop support or a combined system of prop support and roof bolting, nevertheless in the case of heading protected with an individual roof bolting, high speed of separation can cause hazard to the safety of miners.

The paper analyses changes of separation speed in roof rocks, where various types of heading's support were applied. In the analysed workings, i.e. in the heading 2-E1 in the seam 703/1, in the inclined drift Izn in the seam 358/1 and in the inclined drift B-1 in the seam 404/1, the following types of support were applied respectively: the individual roof bolting, the combined system of prop support and roof bolting, and the traditional ŁP-type arch support. Although the workings were located in three different hard coal mines, their location was selected in such a way that the roof conditions were to be comparable in all cases. Due to the fact that it was expected that the rock mass movements would be happening in a long period of time, the workings were drilled in the areas where mining exploitation had not been carried out before.

2. CHARACTERISTICS OF MINING-GEOLOGICAL CONDITIONS AND SUPPORT SCHEMES

2.1. *Rise gallery I-E1 in the seam 703/1 – individual roof bolting*

The area of the seam 703/1, where the rise gallery I-E1 was drilled, is located at the depth of approximately 970 m. The thickness of the seam is 1.9–2.2 m with the inclination of up to 5°. The roof and the floor of the working consist of shale clays and arenaceous shales. The rise gallery I-E1 is in the neighbourhood of the wall's caving goaf. It may be also assumed that the working is not subjectted to the interference of exploitation edges as their vertical distance considerably exceeds 200 m.

The rise gallery I-E1 in the seam 703/1 has a rectangular cross section with the dimensions: width 5.0 m and height 2.5 m. In the heading's roof, seven steel bolts with the total length of 2.35 m were fixed in a row (Fig. 2.1). The distance between the bolts in the row is 0.8 m and the distance between the rows is 0.9 m. The wall of the heading from the side of unmined coal is protected with three rows of steel bolts with the total length of 2.35 m, but from the side of the front, the wall was protected with two rows of workable bolts with the total length of 2.0 m. The length of the section of the heading I-E1, in which individual roof bolting was applied, is 182 m.

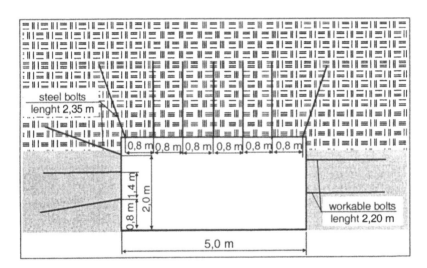

Figure 2.1. Scheme of bolt distribution in the rise gallery I-E1 in the seam 703/1

2.2. *Inclined drift Izn 358/1 – combined system of prop support and roof bolting*

The seam 358/1 in the area of the inclined drift Izn is located at the depth of approximately 900 m. The thickness of the seam is approximately 2.5 m with the inclination of up to 5°. The 1.5-meter thick layer of shale clay constitutes the immediate roof, above which there are sandstones and arenaceous shale. The neighbouring seams were not exploited and the inclined drift Izn was drilled parallel to the caving goaf located in the distance of 50–70 m.

The inclined drift Izn in the seam 358/1 was constructed in the ŁP9/V29-type support, so the width of the working is 5.0 m with the height of 3.5 m (Fig. 2.2). The spacing of the sets in the analysed working is 1.5 m. The support is reinforced with two pairs of bolts fixed to the roof-bar arch at the distance of approximately 1.0 m from the working's axis. The applied bolts have a total length of 2.5 m and were fixed on all the length. The section of the inclined drift Izn subjected to measurements had a length of approximately 100 m.

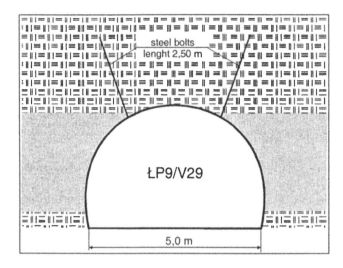

Figure 2.2. Scheme of bolt distribution in the inclined drift Izn in the seam 358/1

2.3. *The inclined drift B-1 in the seam 404/1 – prop support*

The seam 404/1 in the area of the inclined drift B-1 has the thickness of 2.05–2.15 m and the inclination of approximately 5°. In the seam there is an interlayer of shale with coal laminas with the thickness of 0.35–0.40 m. There are no exploitation edges in the premises of the working. However, in its neighbourhood there used to be active wall fields in the seam 403/1 and 404/2. The distance between the seam 404/1 and the seam 403/1 in the area of the inclined drift B-1 is between approximately 10 m and 20 m.

In the immediate roof of the working there is shale clay with occasional interlayers of arenaceous shale and sandstone. In the floor, to approximately 25 m, shale clay with occasional interlayers of arenaceous shale can be found. Immediately below the working, carbonaceous shale or shale clay with coal laminas may occur.

No tectonic disturbance occurs in the runway of the inclined drift B-1.

The working was built in the ŁP-V29/9-type support, so it has a length of 5 m and the height of 3.5 m, with the spacing of 1 m.

3. ANALYSIS OF SEPARATION SPEED

The measurements of separation were carried out using telltales of high and low separation, multi-level telltales and extensometric probes.

In the rise gallery of the wall I-E1, 11 stations for measuring low and high separations as well as multilevel telltales were fixed. The distance between measurement stations is approximately 20 m. The high telltales were fixed at the height of 4.5 m and the low telltales – 2.3 m. The critical value of separation in the rise gallery I-E1 was determined as follows: for low separation – 25 mm, for high separation – 40 mm. The frequency of measurements was between 2 and 4 times per week during the drilling of the rise gallery, and between 5 and 8 times per month after terminating the drilling process.

Five stations for measuring low and high separation were fixed in the inclined drift Izn. High telltales were fixed at the height of 5.3 m, and low telltales – 2.5 m. The distance between the telltales was approximately 20 m. Due to the application of a combined system of prop support and roof bolting, the critical value of separation was not assumed. The frequency of measurements was once per month in the first year and later it was once per three months.

In the inclined drift B-1 the measurements of separation were carried out with the use of extensometric probes. They were fixed in the roof of the working in the two measurement stations, approximately 67 m distant from one another. Roof separations were controlled using the above-mentioned instruments on 18 levels of the immediate and the proper roof to the depth of approximately 7.5 m. The precision of the record, based on the electromagnetic signal, was 0.025 mm. The measurements were initially carried out every week and later once per month.

3.1. *Rise gallery of the wall I-E1*

The measurements of the rise gallery I-E1 were carried out for a period of 22 months. For the sake of the analysis of separation speed, the low and high telltales No 7 and No 8 as well as the multilevel telltales No 3 were used. Fig. 3.1 presents the separation speed recorded with the use the telltales No 7. The RN symbol refers to low separation, whereas the RW symbol refers to high separation.

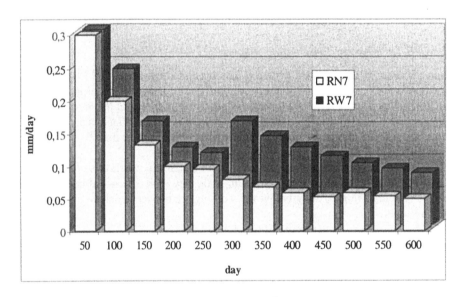

Figure 3.1. Speed of separation in time in the telltales No 7 in the rise gallery I-E1

It may be assumed from the graphs presented that the highest separation speed occurs in the period of approximately 100 days and it reaches nearly 0.3 mm/day. It should be pointed out here that in the period of the initial 50 days of measurement, separation speed was similar for low and high separation. In the consecutive measurements it was observed that separation speed was decreasing systematically, reaching the value of 0.05 mm/day in the final period. The decrease of the analysed parameter could be also observed in relation to high telltales, although in the period between 250th and 300th day of measurement separation speed increased from 0.112 mm/day to 0.16 mm/day. However, this phenomenon occurred only in the high telltales and after this a systematic decrease of separation speed could be observed, and its final value was 0.08 mm/day.

A slightly different character of changes of separation speed in particular periods of time could be observed with the use of the telltales No 8 (Fig. 3.2).

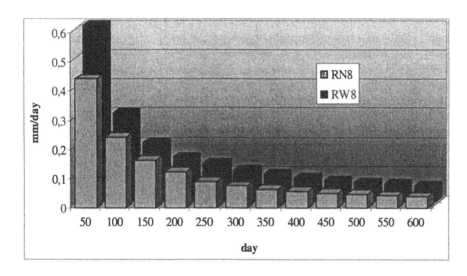

Figure 3.2. Speed of separation in time in the telltales No 8 in the rise gallery I-E1

On the basis of the above graph it may be assumed that separation speed was decreasing systematically, initially having its peak value of 0.44 mm/day for the low telltales and 0.60 mm/day for the high telltales. Increased value of the analysed parameter can be also observed for high separation, which suggests more considerable increase of separation. The final values of separation speed were slightly lower than the ones recorded in the station No 7 and were 0.036 mm/day and 0.053 mm/day respectively.

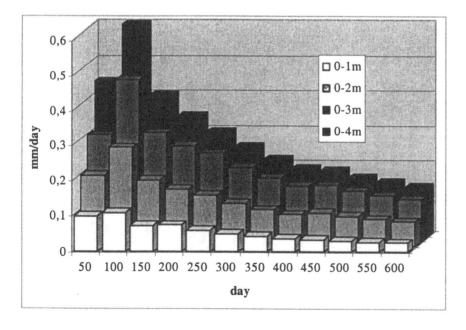

Figure 3.3. Speed of separation in time in the multilevel telltales No 3 in the rise gallery I-E1

The graph representing separation speed for the multilevel telltales seems to offer some interesting possibilities for interpretation (Fig. 3.3). After studying it carefully it may be assumed that the dynamics of separation on particular levels increases significantly from 0.1 mm/day for the level 0–1 m to 0.44 mm/day for the level 0–4 m. Even more significant differences, more than quintuple, occur in the period between the 50th and 100th day of measurement. Then the separation speed reaches 0.6 mm/day for the separation in the packet 0–4 m.

After this period of time the dynamics of changes decreases systematically for all levels, and the differences in values for particular levels, which had occurred in the initial phase of measurements, remained unchanged till the very end of the measurements.

In the case of the multilevel telltales it may be assumed that separation speed depends on an analysed rock thickness. Usually, the thicker the analysed packet of roof rocks, the bigger the total separations. Similarly, using a similar way of monitoring, separation speed of particular sections of the working's roof, which is a difference of separations speed on the analysed lengths, can be determined. For instance, for the rise gallery I-E1 in the initial 50 days of measurement, the separation speed between the second and the third metre of the roof is 0.1 mm/day.

To sum up, it may be claimed that the dynamics of roof rock separation in the rise gallery I-E1 in particular telltales has a very similar course. The biggest values of separation were recorded in the initial period of measurements and, with time, the values were systematically decreasing. As far as the maximal absolute values are concerned, separation speed changed from 0.3 mm/day to 0.6 mm/day depending on the location of the measuring instrument.

3.2. *Inclined drift Izn*

The results from the inclined drift Izn were obtained during a 37-month period of measurements and the low and high telltales No 6 and No 9 were used in order to analyse the separation speed. Figure 3.4 presents the separation speed recorded by the telltales No 6. The symbol RN refers to low separation and RW to high separation.

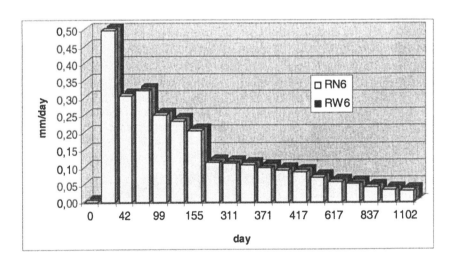

Figure 3.4. Speed of separation in time in the telltales No 6 in the inclined drift Izn

It may be assumed from the presented graphs that the highest sped of separation occurs in the period of the initial 10 days and it is 0.5 mm/day. Until the 155th measurement day, a significantly high value of separation speed occurs and it is 0.2 mm/day. In the later period of time the dynamics of separation speed decreases systematically to the point in which the values of low and high separations are similar. It reflects the changes going on in the packet of the immediate roof at the depth of approximately 2.5 m.

Significantly higher values of separation speed were observed in the case of the telltales No 9 (Fig. 3.5), as in this case separation speed in the period of the initial 10 days was 1.2 mm/day, both for low and high separation. Almost two times lower value of the analysed parameter was obtained in the following 30 measurement days. Also in the case of the telltales No 9, a regular decrease of separation speed occurs and the final values were 0.044 mm/day for high separation and 0.043 mm/day for low separation.

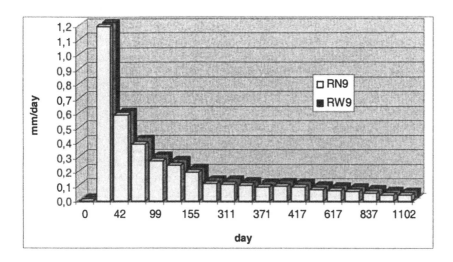

Figure 3.5. Speed of separation in time in the telltales No 9 in the inclined drift Izn

On the basis of the comparison of the obtained results from the ones obtained in the rise gallery I-E1 built in an individual roof bolting, it may be concluded that in the case of the working with the combined system of prop support and roof bolting, separation occurs much faster. The time, in which extreme values are reached, is another factor differentiating the changes of separation speed; for the working with a combined system of prop support and roof bolting they occur in a shorter period of time. However, some similarities can be observed in the character of changes of the analysed parameter as separation speed decreases with time regardless of the type of support system.

3.3. Inclined drift B-1

The measurements of separation in the inclined drift B-1 were carried out for the period of 28 months. This paper presents the results recorded with the use of the probe No 2. Out of the set of 18 results, only the results recorded in 8 selected sensors fixed at the heights of 0.59 m, 1.26 m, 2.30 m, 3.43 m, 4.79 m, 5.82 m, 6.92 m and 7.52 m were used for the sake of the presentation in this paper (Fig. 3.6).

On the basis of these results it may be assumed that separation speed, similarly to the above-mentioned cases, decreases with time, but its character of changes is different. In the period of the initial 7 days, separation speed has a peak value of between 0.26 mm/day and 0.68 mm/day. In the following periods of time separation speed decreases. The graph representing separation speed has a form of a wavy curve, as its values increase and decrease alternately, however with the tendency towards zero. After 171 days of measurement, separation speed of roof rocks does not exceed 0.07 mm/day, and after 444 days – 0.003 mm/day. Despite practically stopping the process of separation until the 594th measurement day, after the 600th day the increase of fissure opening occurs again. As the strata control in the working was maintained by means of a prop support, whose work characteristics is based on cyclic slides alongside with the progressing pressure of the rock mass on the support, the oscillatory movement of roof layers can be caused by the work of the support.

It should be also pointed out that separation speed of roof rocks might be negative, which means the compaction of rock layers and closing fissures. Such a situation may occur not only in the first metres of the roof (sensor at the depth of 0.59 m) but also in the further distance from the working's contour (sensor at the depth of 4.79 m). Due to a very high precision of the measurement and a high density of distribution of measurement sensors, such a result could be obtained only in the case of controlling rock mass movement with the help of the extensometric probe.

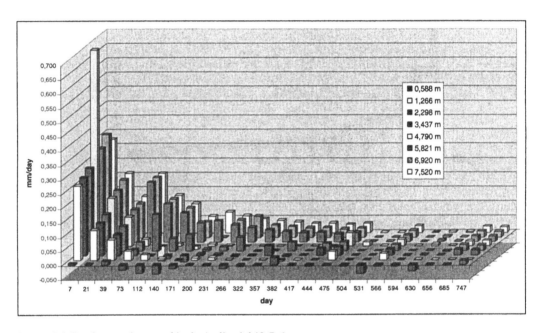

Figure 3.6. Roof separation speed in the inclined drift B-1

4. CONCLUSIONS

On the basis of the above considerations and evaluations of separation speed in underground headings in relation to a type of support system applied, the following conclusions may be drawn:
1. Separation speed always asymptotically decreases to zero, having a value of between 0.3 mm/day and 1.2 mm/day in the initial period of time.
2. The process of disappearance of separation in the case of applying an individual roof bolting is the longest. A rock beam created by fixing the bolts subsides for a much longer period of time and even after 600 days of measurements the speed of dislocations may still have the value of 0.15 mm/day.
3. Roof separation has a changeable character, i.e. it can increase or decrease in the consequent periods of time.

4. Separation speed of roof rocks can be negative, which suggests compaction of rock layers and closing fissures.
5. Only extensometric probes allow for a precise control of roof strata above a working. In this case, the precision of measurement is 0.001 inch. Controlling roof layers with the use of tell-tales does not provide such possibilities.

REFERENCES

Butra J. & Orzepowski S. 2001: Ocena stateczności stropu wyrobiska na podstawie prędkości jego rozwarstwienia. Miesięcznik WUG, nr 5, pp. 22–30.

Butra J. 1998: Kierunki rozwoju metod monitoringu stropu w warunkach wzmożonych ciśnień i przemieszczeń górotworu. Cuprum, nr 3, pp. 77–96.

Fabich S., Pytel W. & Skopiak L. 2000: Badania zachowania się obudowy kotwowej stropu wyrobisk na podstawie pomiarów wielopunktowymi sondami ekstensometrycznymi i kotwami oprzyrządowanymi. Cuprum, nr 17, pp. 73–94.

Fabjanczyk M. 2001: The Use of Monitoring for the Design and Management of Bolting System. Nowoczesne technologie górnicze 2001 – Kotwienie, pp. 135–148.

Gale W.J. & Blackwood R.L. 1987: Stress Distributions and Rock Failure around Coal Mine Roadways. Int. Journal Rock Mech. Min. Scien. & Geomech., Vol. 24, nr 3, pp. 165–173.

Majcherczyk T. & Niedbalski Z. 2002a: Badania zasięgu strefy spękań wokół wyrobiska korytarzowego. Przegląd Górniczy, nr 2, pp. 9–14.

Majcherczyk T. & Niedbalski Z. 2002b: Ocena obudowy podporowo-kotwiowej na podstawie wybranych badań in situ. Przegląd Górniczy, nr 12, pp. 1–7.

Majcherczyk T., Małkowski P. & Niedbalski Z. 2005: Separation of Roof Rock Observed in Headings under Development. Proceedings of the 6th Int. Mining Forum 2005, 23–27 February, Cracow, Szczyrk, Wieliczka. Rotterdam, Balkema, pp. 53–62.

Szymiczek W., Ficek J. & Sobik M. 2001: Nowoczesne metody monitoringu pracy obudowy kotwiowej. Nowoczesne technologie górnicze 2001 – Kotwienie. Gliwice, Ustroń, pp. 453–468.

International Mining Forum 2006, Sobczyk & Kicki (eds) © 2006 Taylor & Francis Group, London, ISBN 0415 401178

The Use of Pre-stressed Tendons in the Polish Coal Mining Industry

Zbigniew Rak, Jerzy Stasica, Michał Stopyra
AGH – University of Science and Technology, Cracow. Poland

ABSTRACT: The paper outlines the practice of roof support with the use of long string wire cable-bolts in Polish coal mines from its beginning till now. Shown are the results of the observations conducted to determine the interaction between the cables and the rockmass in various geological and mining conditions. The documented cases of cablebolt support installation and performance monitoring have been done on such mines as: "Bogdanka", "Sobieski Jaworzno III", "Murcki", "Wieczorek", "Bielszowice", "Piast". The authors also present the results of laboratory research conducted at a suitably prepared test stand as well as new cable construction designs together with their technical characteristics. Further on described are the most effective methods to install cable-bolts in various rocks, characterised by diverse structural and strength properties.

KEYWORDS: Pre-stressed tendons, coal mining, mine tunnels, support reinforcement

1. INTRODUCTION

The beginnings of the use of rockbolts for rock reinforcement dates back to the end of the 19th century. On a wider scale, however, this support method was employed only half a century later. As from the 1950s rockbolting became the basic means of support used in Europe in underground mining and civil engineering excavations. The current world usage or rock anchors is estimated at some tens of millions of units installed per year.

Rope tendons, an important type of rock anchors, have been in use in the mining industry for more than thirty years. They were first installed during the field trials conducted in underground metal ore mines at the late 1960s. Grouted rope strands 15.2 mm in diameter were used in the 1970s as rock reinforcements. At that time they did not yet come with auxiliary fittings such as faceplates and locking nuts. By the end of the 1980s rope bolts were already in common use in underground coal mines of South Africa, Australia and the USA.

Thanks to their construction rope bolts possess several advantages making them particularly useful in the most difficult mining and geological conditions, particularly conditions of high static stress and large de-stressed zones around excavations. In high seismic hazard areas they perform the function of so called "barrier support".

Cablebolts have been in wider use in the Polish coal mining industry for the past 10 years only. Extensive development has taken place during this period of time, both in the design of cable construction and in the technology of longhole bolting itself. Tendon construction, technology and mechanization of installation and support performance monitoring – development took place in all these areas. The paper presents the practical experience gained in Polish mines in the use of longhole rope and string wire cablebolt support so far. The authors also present the results of laboratory research conducted on this support type and present new cable anchor designs together with the most important technical characteristics of the described units. Further on described are the most effective methods to install wire cablebolts in rocks characterised by diverse structural and strength properties.

2. "HIGH" BOLTING WITH THE USE OF STRING WIRE AND ROPE CABLEBOLTS

It is often necessary in underground mines to install, either as primary or secondary support, rockbolts whose lengths exceed (sometimes manifold) the height of the excavation. Traditionally the Polish nomenclature would call rockbolting "high" whenever the length of support units exceeded the height of the supported excavation. However, in the case of large-sized excavations this definition does not convey the true meaning of high bolting. Many authors formulated their own definitions, stipulating different rockbolt lengths to differentiate between "high" and "low" bolting. It is largely accepted in the Polish coal mining industry to refer to rockbolting as "high" in the cases when anchor length exceeds 3,0 metres.

It is relatively seldom that stiff steel rods are used for high bolting. The reason is the necessity to join rod sections (usually with threaded joins), making the installation process troublesome. Larger diameter holes are normally needed to accommodate them as well. The notch effect of the joint weakens the whole unit. For these reasons cables rather than rods are utilized. The most popular definition of a cablebolt describes its construction as comprising several individual or twisted steel wires (anchors suitable for pre-tensioned installations) or twisted strand ropes [Stilborg 1994]. In the Polish nomenclature these two types are called respectively string wire and rope cablebolts.

The most important feature of a cablebolt is its flexibility, rendering it possible to install cable of practically any length regardless the dimensions of the excavation. This makes it possible to reinforce a wider zone of rock around an excavation than can be done with "low" bolting. The important advantage of cable bolts over plain rods is their notably greater tensile strength for the same diameter tendons. Subjected to the same loads cables normally display smaller deformations. However, it is possible to increase yielding capacity of a cable without changes to its construction. Thanks to the named advantages, cablebolts are used increasingly more frequently for "low" rockbolting as well [Rataj 1995, Rataj 2000].

Figure 2.1 shows types of string wire cablebolts most commonly used in the world mining industry. Of them, only the types shown as a, d and f are used on a larger scale in Polish coal mines. In the 1990s several field trials with 18 and 20 mm in diameter rope cablebolts were conducted. An example of such construction is a unit manufactured and marketed as IR-3 by INTERRAMM. It is manufactured from standard 18 mm diameter twisted strand ropes type $6\times7+A_0$ (18,0-$6\times7+A_0$-Z/s-n-II-g-140) (Photo 2.1a, b and c). The tendon of this rope comprises 6 strands, each of them consisting of 7 twisted steel wires. Constructed in this way rope comprises 42 steel wires and a soft core, usually made from jute fibres. The cross-sectional area of steel wires in a 18 mm diameter rope is approximately 131 mm^2. Its breaking force is at least 160 kN. The tendon of this cablebolt is modified by untwisting the original rope over the length corresponding to the length of bond. The rope prepared in this way and grouted in a hole forms a kind of disorderly steel reinforcement.

The maximum diameter of the bulbed part of a cable called "cage" is up to 38 mm. In its thinnest part, where the wires remain twisted the diameter of the tendon equals to approximately 24 mm. The advantage of this construction is that the cable remains securely fastened even if the grout reveals certain contraction. Moreover, contrary to smooth cable, after grouting the bulbed strands do not have the tendency to break loose from the grout under load. The locking mechanism of this bolt comprises the 3-part Gifford lock.

Figure 2.2 shows an example of a load-deformation graph of IR-3 cable bolt. The units used for the test were left ungrouted over a length of 1,0 metre. The measured load-bearing capacity was not less than 175 kN and the total elongation of the tendon at the moment of failure was approx. 5%. Cablebolts of this type were frequently used as reinforcement in face gates supported solely by steel dowels at, among others, KWK "Ziemowit", KWK "Wesoła" and KWK "Czeczott" mines. They were also used at road crossings at KWK "Piast" and KWK "Wesoła" and to reinforce footwall at KWK "Wesoła" mine. Despite promising laboratory test results their use have been discontinued due to inability to overcome technological problems, such as difficulties with correct installation of the cablebolt's locking mechanism (Gifford lock) in practical underground working conditions and problems with forcing the cables into boreholes pre-filled with thick cement grout. It was also in the 1990s when the use of string wire cablebolts designed for pre-stressed installations began in Poland on a larger scale. They were 15,5 mm in diameter, which made them easier to insert into boreholes. Cablebolts of this type are manufactured by ARNALL POLAND and are known under market name KL-15,5 or KL-15,5K (Photo 2.2a, b and c).

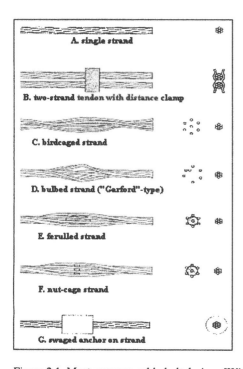

A. single strand

B. two-strand tendon with distance clamp

C. birdcaged strand

D. bulbed strand ("Garford"-type)

E. ferulled strand

F. nut-cage strand

G. swaged anchor on strand

Figure 2.1. Most common cable bolt designs [Windsor 1992]

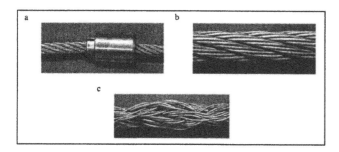

a   b

c

Photo 2.1. IR-3 rope cablebolt: a – lock, b – cable, c – untwisted length of tendon (birdcage)

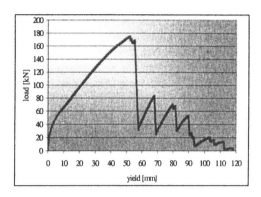

Figure 2.2. Working characteristics of IR-4 cablebolt

The cable of this bolt, manufactured according to PN-71/M-80236 standard, comprises seven axially twisted steel wires. The total cross-sectional area of the wires is 143 mm². The breaking force of such tendon is approximately 250 kN. The cable may be either smooth (KL-15,5) or geometrically modified (KL-15,5K) by the formation of small cages, about 100 mm long and 26 to 28 mm in diameter. The cages are usually spaced every 200 mm. Cablebolts with greater birdcage spacings are also manufactured. The locking mechanism designed for this bolt is also the 3-part Gifford lock.

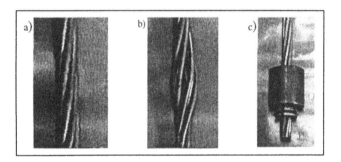

Photo 2.2. KL-15,5 (KL-15,5K) wire cablebolt: a – cable, b – birdcage, c – lock

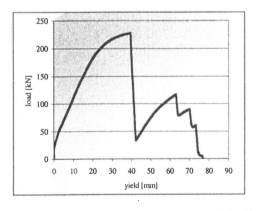

Figure 2.3. Working characteristics of KL 15,5 cablebolt

The stress-strain characteristics of KL 15,5 cablebolt is shown on picture 2.3. The total elongation of the bolt at the moment of failure was approx. 4%.

Another wire cablebolt commonly used in Polish coal mines is IR-4, manufactured by INTER-RAMM BIS (Photo 2.3a, b and c). The cable of this bolt comprises eight 6 mm diameter bound in parallel wires.

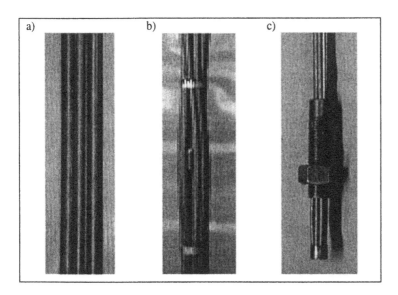

Photo 2.3. IR-4 wire cablebolt: a – cable, b – birdcage, c – lock

The total cross-sectional area of the wires is approximately 226.2 mm². The breaking force of such cable is approximately 320 kN, with the manufacturer-guaranteed load-bearing capacity of 280 kN. One meter of the cable weighs 1.75 kg. The wires comprising the cable are made of BII spring wire as per PN-71/M-80057 standard. The cable of the IR-4A cablebolt may be either smooth or modified to form birdcages (Fig. 2.3b). The diameter of the smooth cable is 22–24 mm. In the case of the modified one, its maximum diameter is 38 mm. The lock of IR-4 cablebolt comprises a threaded sleeve with a nut. The cable is secured in the sleeve by means of a steel wedge. At load, individual wires of the cable move and gradually wedge themselves tight in the sleeve until their further movement is stopped. Such design provides for proper wedging of the wires in the lock and allows eliminating their movement relative to each other already at installation, when initial load is applied. Wedging of the wires at an early stage of pre-tensioning is important for the cable to work properly. The strain-stress characteristics of IR-4 wire cablebolt is shown on Figure 2.4. The total movement of the cable at failure, which is the sum of wire movement on the lock and the cable elongation, does not exceed 5%. Based on the IR-4 design new types of cablebolts similar in construction were created – IR-4C and IR-4W. IR-4C cablebolt was designed jointly by the scientists from the Underground Mining Department of AGH (the Technical University of Mining and Metallurgy, Kraków) and INTERRAMM BIS. The cable of this bolt comprises nine twisted 6 mm diameter each wires. The total cross-sectional area of the wires is approximately 254.34 mm². The breaking force of the cable is not less than 350 kN, and the elongation at failure does not exceed 5% (Figure 2.5). The diameter of IR-4C cable is between 22 and 24 mm.

The cable of IR-4W bolt comprises seven 8 mm diameter wires. The cablebolt is about 26 mm in diameter.

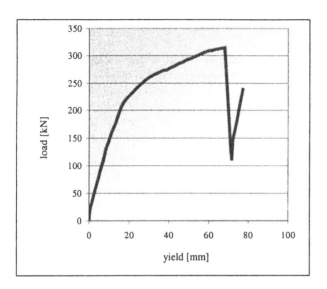

Fig. 2.4. The strain-stress characteristics of IR-4 wire cablebolt

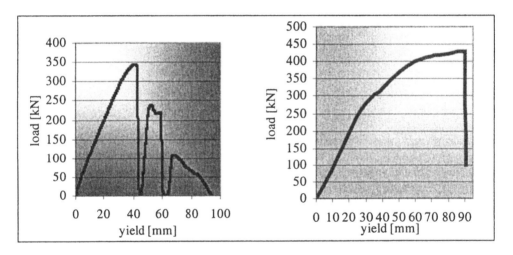

Figure 2.5 (left). Working characteristics of IR-4C cablebolt
Figure 2.6 (right). Working characteristics of IR-4W cablebolt

The total cross-sectional area of the wires is approximately 351.68 mm^2. The outside diameter of the sleeve forming the lock has in this case been increased to 39 mm. It is made from the same material, i.e. 40 HM steel (Cr-Mo steel, Polish classification). The maximum load at failure for this bolt is approximately 440 kN, with the elongation not exceeding 5% – Figure 2.6. Because of its relatively good strength characteristics (highest load-bearing capacity of all the cablebolts used in the Polish coal mining industry), IR-4W cablebolts were subjected to dynamic loading tests. The tests were conducted in the specialised laboratory of the Central Mining Institute in Katowice. Bolts were subjected to dynamic loads created by the impact of a falling weight. Kinetic energy of each impact was 25 kJ. Some of the results of the research, presented as load-time graphs, are shown on Figures 2.7 and 2.8.

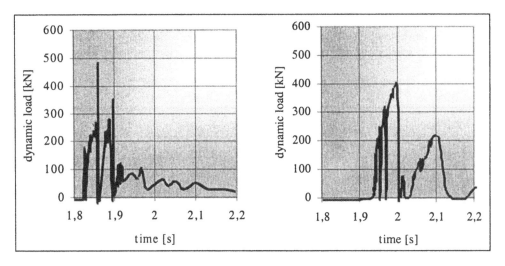

Figure 2.7 (left). Load-time graph for IR-4C cablebolt obtained by impact-loading it with 25 kJ of dynamic energy

Figure 2.8 (right). Load-time graph for IR-4W cablebolt obtained by impact-loading it with 25 kJ of dynamic energy

The obtained results allowed classifying IR-4W string wire cablebolt as suitable for installation in dynamic load conditions.

3. WIRE CABLEBOLT INSTALLATION METHODOLOGIES

The choice of installation technique is determined by several factors, such as type of grout, the length and diameter of the support hole, the grouted length of the hole and the available equipment.

In general, three cablebolt installation methods can be named:
1. With the use of thin cement grouts and liquid resins.
2. With the use of thick cement grouts.
3. With the use of cement or resin capsules.

Until recently the first method was the only one utilized to install long cablebolts in roof and sidewalls. The principle of the method is schematically explained on Figure 3.1. Used in this case particular are thin cementitious grouts or two-component resins. After the hole has been drilled a cablebolt equipped with two tubes is inserted − a breather tube, with length equal to that of the cable, and a shorter tube for injecting the grout. The mouth of the hole is then sealed with a 30 cm long polyurethane foam plug. To the prepared in this way hole liquid grout is injected. The hole is filled starting from the bottom, with the air escaping from the hole via the breather tube. Pumping of the grout is stopped when it starts to come out the breather tube. The major disadvantage of this method is its high labour intensity. Additional problem in the cases where cement grouts are used is their long setting time and the possibility of the grout to penetrate from the hole into fractures. Additional disadvantage are the relatively poor strength parameters of the cement stone formed from the setting grout. The use of two-component resins considerably increases the cost of grouting and does not eliminate the risk of the grout escaping into the cracks. In the second method thickened cement grouts are used (Figure 3.2). In this case filling of the hole starts at its further end. The bolt is inserted by hand (or mechanically) into the hole, which is completely or partially filled with grout. The method allows achieving higher bond strengths, mainly thanks to the relatively high strength parameters of grout (even above 50 MPa). With the lower water-cement ratio the setting time is also shortened (below 1 h). Troublesome with this method is the necessity to prepare fresh grout directly before application and the labour-consuming process of thorough cleaning of the grouting equipment.

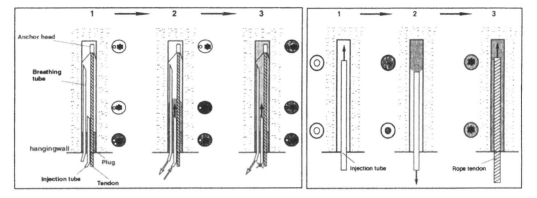

Figure 3.1. Cablebolt installation method with the use of thin cementitious grouts and liquid two-component resins

Figure 3.2. Cablebolt installation with the use of thick cementitious grouts

At the moment, the most often used in Polish mines for cablebolt installation is the method with cement or resin capsules. Thanks to its simplicity and susceptibility to mechanisation the method allows to substantially shortening the time of the cablebolt installation process. This in turn affords installation directly at the face of the driven heading or, in the worst case, just behind the road-header. Bolting of rock surface as soon as possible after it is exposed is beneficial to the stability of the excavation. The use of fast-setting resin capsules allows to pre-tension the bolt directly after the installation, resulting in its almost instantaneous interaction with the rockmass. The disadvantage of the method is impossibility of grouting the cable over length greater than 1,5 metre. Usually two 60 cm long 3-minute cartridges per bolt are used due to the flexibility of the cable and the resistance encountered while mixing the capsules. It must be noted that both laboratory and in situ tests show that cable failure loads can be achieved for bond lengths as small as 50 cm.

One of the most important parameters determining the proper installation of cablebolts with resin capsules is the torque available on the installation equipment. The most common type of equipment used with this method is a pneumatic roofbolter (e.g. Gopher). However if pressure of the compressed air in the supply line is too low, it does not guarantee correct bolt installation. Relatively recently introduced hydraulic roofbolters VSP-1 powered by their own hydraulic aggregate were introduced. High efficiency of this equipment resulted in substantial increase in the quality of bolt installation. And awakened interest in "high" rockbolting as well. In average rock conditions the units allow to install up to ten 6-metre-long wire cablebolts per shift.

4. APPLICATIONS OF STRING WIRE CABLEBOLTS IN POLISH COAL MINES

The use of long bolting is relatively widespread in the Polish coal mining industry.
Some of the possible applications for string wire cablebolts are named below:
– As reinforcement for tunnel steel set support, to eliminate the need for additional props. This is particularly important at face-gate intersections, where no space for additional props is available.

- To reinforce tunnel support in areas of increased stress related to geological disturbances and mining.
- To reinforce set support of a face gate maintained along the bog-side of a longwall.
- To reinforce tunnel support at tunnel crossings.
- To reinforce cross-gate support to facilitate equipping and starting of a new longwall.
- To reinforce support of large-size excavations.
- To reinforce support of capital excavations with expected long service life.
- To reinforce tunnel support to facilitate the installation of overhead rail transport ways.
- As a preventive measure in excavations located in areas with high seismic hazard.
- As reinforcement in tunnels supported solely by rockbolt support.
- To reinforce rockmass around mining excavations.

Concentration of extraction, high face advance rates and increasingly difficult geological conditions in the mined coalbeds force engineers to look for new ways to support face-gate intersections. Until recently the role of support reinforcement was played by short steel rods used to anchor the caps of the installed steel sets. It happened frequently, however, that the fracture zone around the excavation was developed to such an extent that the length of the used rods was insufficient. For several years now some mines, among them ZG "Sobieski", LW "Bogdanka" and KWK "Murcki", use string wire cablebolts to support longwall-gate crossings and do away with prop support in these areas. Anchoring or fastening of arch segments is usually done with the use of additional V-section horizontal members of various lengths installed along the axis of the tunnel. Anchors are installed through the especially for the purpose burned holes (Figure 4.1). The length of the bracing and the spacing between the anchors depend on local mining and geological conditions as well as the distance of the place of the installation from the face. Installation directly at the face requires using short braces, extending only between two sets. This limitation does not apply where installation is done further away from the face. Bracing with lengths up to 6 metres may then be used.

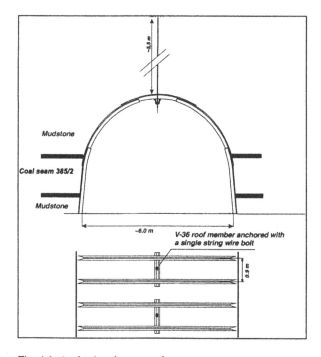

Fig. 4.1. Anchoring the sets at face-gate

The described method allows maintaining longwall-gate intersections without resorting to the use of additional prop support. Combined with a solid backfill rib it dramatically improves conditions in gates located on the bog side of the longwalls (ZG "Sobieski" and KWK "Murcki" mines), (Figure 4.2). Increased interest in high bolting to support cross-gates is recently observed (among others LW "Bogdanka", KWK "Murcki", KWK "Piast", KWK "Polska-Wirek", KWK "Bielszo-wice" mines). These excavations are characterised by extreme widths. No roof set members are able to resist the loads exerted on them by roof rock without additional reinforcement.

The surface of the exposed roof of such excavations is frequently flat (for example where ŁPr, ŁPro set types are used) and impossible to hold without additional support. The traditional method, utilizing a line of props installed along the middle axis of the tunnel, hinders the subsequent longwall equipping process.

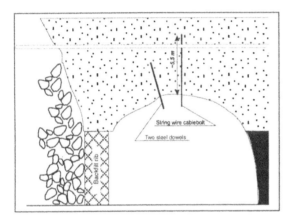

Fig. 4.2. The face-gate maintained behind the face

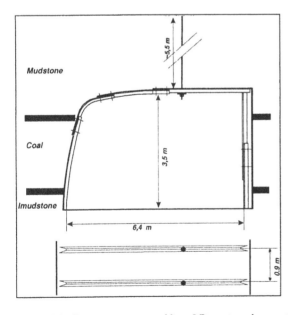

Figure 4.3. Cross-gate supported by a ŁPro set-anchor system

58

Replacing props with cablebolts is effective and improves comfort and safety of work of the equipping crew. Anchoring roof members allows also reclaiming the used props and side members of the installed arches.

Examples of anchor-set support of cross-gates are shown on Figures 4.3 and 4.4.

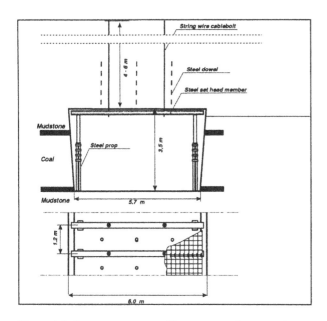

Figure 4.4. Cross-gate supported by a rectangular set-anchor system

Figure 4.5 shows schematically the moment of the newly-installed mechanized longwall supports stepping out of the cross-gate supported by rectangular sets and cablebolts. In this case the set members are stripped successively after the loading of each mechanized support unit. Additional advantage of the method is the flat roof of the excavation, which eliminates the need to use cribbing to fill the void between the support's roof shield and the roof.

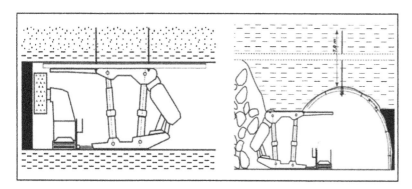

Figure 4.5 (left). Mechanized support units stepping out from cross-gate supported by rectangular sets and cablebolts
Figure 4.6 (right). Longwall traversing a tunnel

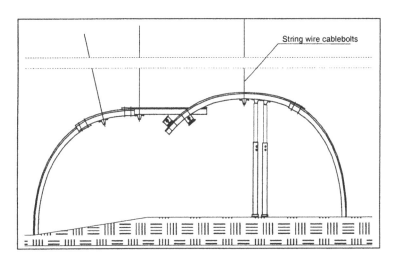

Figure 4.7. The method of tunnel crossing support used at ZG "Sobieski" mine

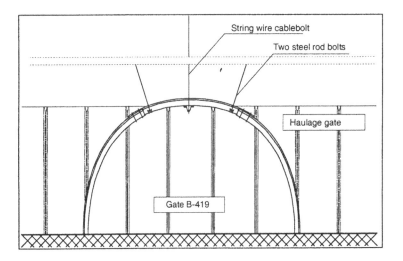

Figure 4.8. The method of tunnel crossing support used at KWK "Murcki" mine

"High" rockbolting serves similar purpose when longwall face traverses a tunnel or enters the cross-gate developed at the furthermost limit where the longwall equipment is to be stripped (Figure 4.6). In this case the extent of the fractured zone around the excavation is so wide that only the use of long bolts can guarantee their stability.

For the same reasons string wire cablebolts may be used to reinforce support installed at tunnel crossings and break-ways. Examples of bolting patterns at tunnel crossings are schematically shown on Figures 4.7 and 4.8 (KWK "Murcki" and ZG "Sobieski" mines). Figure 4.9 shows a method to support a two-level crossing developed at KWK "Murcki" mine. In all the presented examples cablebolts were used mainly to eliminate or substantially limit the number of prop supports, which increased the working space for the crew and transport and haulage equipment. Cablebolts installed in the roof substantially limit the extend of the fracture zone and increase the life of the excavations reinforced in this way.

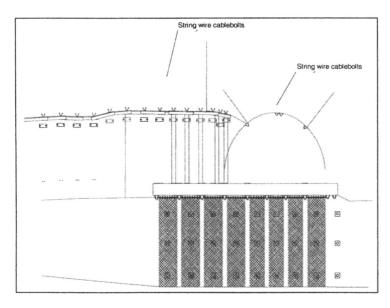

Figure 4.9. The method of support of a two-level crossing at KWK "Murcki" mine

5. FINAL REMARKS

The practical examples of utilization of long bolting presented in this paper certainly do not cover all the aspects related to this topic. We are aware that the mining industry creates opportunity for many more applications for string wire and rope cablebolts. One of such applications, at the moment not yet experimented with in Poland, seems to be the use of cablebolts as barrier support in conditions of high seismic hazard. The research currently under way seems to show that high bolting reduces the intensity of seismicity, hence increasing work safety of the crew. The authors hope that this paper, by summarising the achievements in and applications of high bolting, will serve to promote this support system and stem research into new designs and applications.

REFERENCES

Rataj M. 1995: Poszukiwanie nowych rozwiązań obudowy kotwiowej na przykładzie firmy ANI Arnall, największego producenta obudów kotwiowych w Australii. Materiały Szkoły Eksploatacji Podziemnej '95, str. 165–172.

Rataj M. 1998: Zastosowanie kotwi linowych w australijskich kopalniach węgla. Materiały konferencyjne – II Konferencja pt. "Obudowa kotwiowa jako skuteczny sposób zabezpieczania wyrobisk górniczych", Świeradów, str. 175.

Stilborg B. 1994: Professionals Users Handbook for Rock Bolting. Series on Rock and Soil Mechanics, Trans-Tech Publications.

Windsor C.R. 1992: Cable Bolting for Underground and Surface Excavations. Rock Support, (Kaiser P.K. and McCreath D. eds), Rotterdam: A.A. Balkema, pp. 349–376.

International Mining Forum 2006, Sobczyk & Kicki (eds) © 2006 Taylor & Francis Group, London, ISBN 0415 401178

Contribution to the Prediction of Ground Surface Movements Caused by a Rising Water Level in a Flooded Mine

Anton Sroka
Technische Universität Bergakademie Freiberg. Freiberg, Germany

ABSTRACT: Survey measurements indicate that ceasing to pump mine water out of closed coal mines results in surface heave. In the practice of the German mine surveying the mathematical basis for the projection of such ground movements are the formulae derived by Pöttgens (1985) and Fenk (2000).

The paper critically evaluates these methods and proposes modifications from the rock engineering and the science of mining damages points of view.

1. INTRODUCTION

Ceasing to pump water from closed underground coal mines usually results in the mine water level rising considerably over vast expanse. The research conducted in many countries showed that such planned flooding of mines leads to noticeable heave of the ground surface over large areas e.g. [Oberste-Brink 1940, Pöttgens 1985, Fenk 2000, Goerke-Mallet 2000, Baglikow 2003, Heitfeld *et al.* 2004, Mühlenbeck 2005, Sroka 2005]. The question that needs to be asked is whether the upward movement of the previously sinking ground surface may have a negative impact on the surface buildings, i.e. whether it may lead to mine damages.

Heitfeld *et al.* (2004) noticed that during flooding of Sophia-Jakoba mine in the Aachen region a discontinuity formed near the outcrops of Meinweg and Ruhrrand faults. The resultant surface deformation was 8–9 km long and 6–7 cm high. About 110 damages to structures, 9 of them beyond repair, were observed along this line.

The question regarding possible mine damages is closely connected to the question whether it is possible to predict ground movements caused by mine flooding. This presents researchers with a task to describe and analyse the current state of knowledge relating to the aforementioned problem. The obtained results must be then generalised and extrapolated to apply in the specific rock-mass and mining conditions characteristic to the examined coal mining region.

2. THE PÖTTGENS METHOD (1985, 1998)

Pöttgens says that the zone of broken rock formed in the roof of a mined-out coal-seam is characterised by increased porosity and permeability. A rise of the water level in a flooded mine leads to the increase of pressure in the caving zone and results in its expansion due to a decreased effective vertical stress. Subsequently the hydrostatic force acting on the solid strata above the caving zone heaves the rock right up to the surface. Pöttgens claims that the extent of the movements depends on the increase of mine water level, mining depth, seam thickness and the extend of extraction.

The mathematical foundation of the Pöttgens method is the Geerstma solution (1973), dealing with surface subsidence related to natural gas exploitation.

The solution was based on the poroelastic theory and two assumptions:
– Linear relationship between stress and strain, and
– Homogeneity of the whole rock-mass.

The convergence or compaction of the porous gas-bearing strata i.e. decrease in its thickness due to gas extraction resulting in decreasing its pressure, can be calculated after Geertsma from the following formula:

$$\Delta M = c_m \cdot M \cdot \Delta p \qquad (1)$$

where ΔM – deposit compaction [m]; M – deposit thickness [m]; Δp – drop in the pore pressure [Bar], and c_m – compaction coefficient [Bar^{-1}].

Compaction coefficient c_m is equal to the relative change of deposit width at a pressure drop of 1 Bar. Generally speaking its value depends on rock-mass characteristics, porosity, initial pore pressure and deposit depth.

According to Geertsma a unit subsidence of a point on the surface can be defined for an infinitely small unit volume of deposit dV from the following equation:

$$ds\,(r, H) = \frac{c_m\,(1-\upsilon)}{\pi} \cdot \frac{H}{\left(r^2 + H^2\right)^{3/2}} \cdot dV \qquad (2)$$

where ds (r, H) – unit subsidence of a point on the ground surface; υ – Poisson ratio; H – depth of the deposit unit volume, and r – horizontal distance between the point and the deposit unit volume.

The solution is known in literature under the name of „the nucleus of strain concept".

For a circular gas reservoir with radius R the formula to calculate the maximum surface subsidence is as follows:

$$s_{max}\,(R, H) = 2 \cdot (1-\upsilon) \cdot \left[1 - \frac{C}{\sqrt{1+C^2}} \right] \cdot \Delta M = 2 \cdot (1-\upsilon) \cdot \left[1 - \frac{C}{\sqrt{1+C^2}} \right] \cdot c_m \cdot M \cdot \Delta p \qquad (3)$$

where $C = H/R$.

The point of maximum subsidence is located above the centre of the circular gas deposit. The equation (3) was used by Pöttgens as the basis for his method of calculating the ground surface heave caused by a rising level of water flooding a mine.

As already mentioned Pöttgens assumed that an increase of pressure in the zone of broken roof rock leads to its elastic expansion. Such expansion was defined as the opposite to compaction:

$$\Delta h = d_m \cdot h \cdot \Delta p$$

where Δh – broken roof rock zone expansion; h – thickness of the broken roof rock zone; Δp – – pressure increase caused in the zone of broken roof rock by the rising level of water flooding the mine; d_m – expansion coefficient of the broken roof rock.

The vertical and horizontal positions and the extent of the zone of broken roof rock were defined by the following assumption:
– The zone is situated on the exploitation level.
– The height of the zone is equal to four widths of the mined-out coal seam, and
– The horizontal extent of the zone is equal to the extent of the extraction.

The solution describing the maximum heave of a point on the ground surface situated above a circular reservoir (i.e. a circular mining working filled with water) proposed by Pöttgens is analogical to that proposed by Geertsma (3).

It assumes the following form:

$$h_{max}(R,H) = 2 \cdot (1-\upsilon) \cdot \left[1 - \frac{C}{\sqrt{1+C^2}}\right] \cdot \Delta h = 2 \cdot (1-\upsilon) \cdot \left[1 - \frac{C}{\sqrt{1+C^2}}\right] \cdot d_m \cdot h \cdot \Delta p \qquad (4)$$

It is possible to assess the heave of any point on the surface from a graphical representation of the function:

$$y(C) = 1 - \frac{C}{\sqrt{1+C^2}} \qquad (5)$$

In the case of extraction of a larger number of coal seams the principle of linear superposition with a constant value of the broken roof rock expansion coefficient d_m is applied [Wings, Misere, Pöttgens 2004].

Pöttgens estimated the value of the broken roof rock expansion coefficient for the Limburg (Holland) coal mining region from in-situ subsidence and heave measurements.

The value is equal to d_m.

$$d_m = 0{,}35 \cdot 10^{-3} \text{ Bar} = 0{,}35 \cdot 10^{-2} \text{ m}^2/\text{MN}.$$

Some of the measurements are shown in Figure 1.

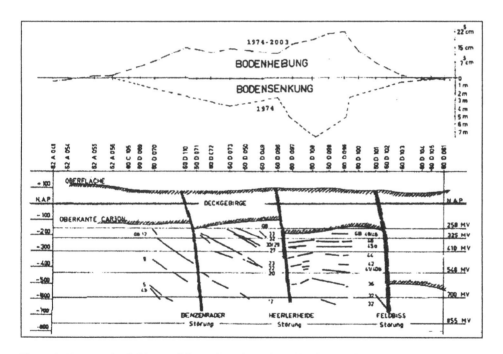

Figure 1. Heave and subsidence of the surface along the Aalbeek-Hoensbroek-Schinveld profile

The above figure shows ground surface heave measurements done in the years 1974–2003 and the mine subsidence calculated for the years 1915–1974.

Pöttgens points out the seemingly evident similarity of both graphs. Moreover, an analysis of the results of the measurements combined with the geological and mining profile showed that not only discontinuous subsidence but also discontinuous heave of the surface is possible in the area of the faults' surface outcrops.

According to Pöttgens this is possible only if extraction at each side of the fault was conducted at different rate and even so only in places where discontinuous subsidence previously occurred.

The relationship between the maximum* subsidence and the maximum* heave may for all practical purposes be presented by the constructed by Pöttgens (1985, 1998) nomograph (Fig. 2) (*the author's interpretation).

The nomograph allows calculating the expected maximum ground surface heave as a percentage of the previous maximum subsidence if the following are known: extraction geometry (GA), mine water geometry (GG) and the mining depth.

According to Wings, Misere and Pöttgens (2004) it is particularly difficult to properly assess GA and GG.

The example shown in Figure 2 pertains to mining depth of 800 m. The extraction geometry is theoretically equivalent to a circle with a radius equal to the radius of influence of the mining (GA = = 100%) and the water horizon will rise right to the surface.

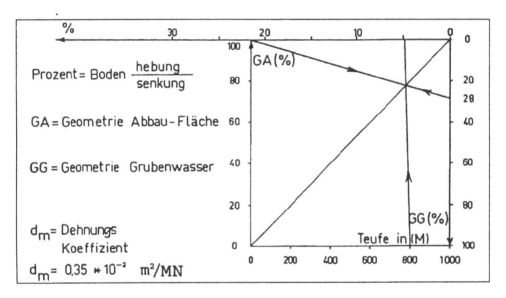

Figure 2. Heave-subsidence relationship

Calculated from the nomograph maximum heave is approximately equal to 5% of the maximum subsidence. According to Pöttgens (1998) the maximum ground surface heave may reach 2–5% of the maximum subsidence if the mine water horizon reaches the ground surface.

Upward movements of this magnitude do not normally cause any damage to surface structures. Pöttgens sets the limit value of deformations, which may result in damages at 0,25 mm/m in compression or tension. This obviously does not pertain to outcrops of faults on the surface and the areas of previous discontinuous deformations.

3. ANALYTICAL METHOD OF FENK (2000)

Similarly to Pöttgens, Fenk based his method on the assumption that a rise in the water level in a flooded mine leads, due to hydrostatic upward pressure, to an elastic expansion of the zone of broken rock situated in the roof of the mined-out coal seam.

The formula for calculating the upward movement of the rock-mass he came up with following the assumption is as follows:

$$h = d_m \cdot \gamma_w \cdot \left[\frac{E_s}{H \cdot \gamma_G} - 1 \right] \cdot W \cdot s_\epsilon \tag{6}$$

where: h – maximum heave of the ground surface, at the moment when the level of water flooding the mine stabilizes [m]; d_m – expansion coefficient of the broken roof rock [m²/MN]; γ_w – specific gravity of water [0,01 MN/m³]; γ_G – specific gravity of the rock-mass [MN/m³]; H – mining depth; E_s – modulus of rigidity of the broken roof rock [MN/m²]; W – increase of the water level in the rockmass [m], and s_ϵ – maximum surface subsidence caused by the extraction [m].

In the case when $W = H$, i.e. the water level rises up to the ground surface, the formula (6) assumes the following form:

$$h = d_m \cdot \gamma_w \cdot \left[\frac{E_s}{\gamma_G} - H \right] \cdot s_\epsilon \tag{7}$$

The analytical solution presented by Fenk was derived for a so-called one seam model. According to the author it only allows calculating the maximum possible heave of the ground surface and contrary to the Pöttgens solution does not require determining the height of the caving zone. Because of that it is impossible to compare d_m coefficients appearing in the Pöttgens's and Fenk's formulae.

Formula (7) has been verified by Fenk (2000) for the Zwickau coal mining region by in-situ measurements of the surface heave.

The verification was based on a regression function, which is the theoretical solution of (7) and has the form:

$$h = a \cdot (b - H) \cdot s_\epsilon .$$

The unknowns a and b were established from the measured values of heave and maximum subsidence and the depth of mining under the 26 measuring stations. Due to the fact that more than one seam was mined in the Zwickau mining region, each time Fenk accepted the mining depth as equal to the depth of the deepest coal seam (!).

After applying the regression analysis Fenk derived the following formula:

$$h = 0,024 \cdot (2024 - H) \cdot s_\epsilon \quad \text{(h w mm; } s_\epsilon, H \text{ w m)} \tag{8}$$

Hence the values of the geo-technical parameters d_m and E_s are equal to:

$$d_m = 0,24 \cdot 10^{-2} \text{ m}^2 / \text{MN},$$

$$E_s = 46,6 \text{ MN}/\text{m}^2 \text{ dla } \gamma_G = 0,023 \text{ MN}/\text{m}^3.$$

The measuring stations were individually assigned corresponding values of mining depths as equal to that directly underneath them. For this reason the measuring stations located outside the mining area were excluded from the assessment, as assigning them any value of mining depth was impossible.

In order to calculate the value of upward movement of the points situated outside the mined-out area Fenk proposed the following empirical formula:

$$h = p \sqrt{s_\epsilon} \tag{9}$$

Specifically for the Zwickau mining region Fenk proposed the following equation based on the data collected there in 1997:

$$h = 58,1 \sqrt{s_\epsilon} \quad \text{(h w mm; } s_\epsilon \text{ w m)} \tag{10}$$

According to Fenk (2000) the advantage of his solution over Pöttgens's is that the analyst does not have to possess the knowledge of the height of the caving zone, the geometry of the conducted mining and the interaction between individual excavations to be able to calculate the extent of heave of the ground surface.

4. CRITICAL REMARKS REGARDING PÖTTGENS'S AND FENK'S METHODS

The methods proposed by Pöttgens and Fenk are the first analytical methods used to predict the amount of ground surface heave caused by water flooding a mine.

Both methods are based on the relationship between the amount of heave and the amount of the previously occurred mining-related surface subsidence. However, in the case of the Pöttgens's method, combining the two is not necessary (see formula 4) and the author alleges that the connection was created purely and somewhat artificially to afford possibility to estimate the expected heave from the earlier subsidence.

It may appear at the first glance that a relationship between the two graphs shown in Figure 1 really exists. A careful analysis shows, however, that the places of maximum subsidence and heave do not correspond. The same applies to other characteristic parameters of the graphs, e.g. the points of their maximum slope.

Creating a ground heave distribution graph from subsidence distribution data will lead to obtaining substantially erroneous, both quantitatively and qualitatively, result.

The recent works of Goerke-Mallet (2000) and Mühlenbeck (2005) contradict the claim that any relationship between the amounts of ground heave and previous subsidence exists.

According to Pöttgens (1985) heave is primarily calculated from formula (4). The graphical integration grid produced by solving the formula allows establishing the value of heave at any point on the surface provided that the input data i.e. the extraction geometry, the coal seam width, the mining depth and the mine water level are known. It is also necessary to know the values of the expansion coefficient d_m and Poisson's ratio υ.

For the cases of multiple coal seam extraction Pöttgens assumed the applicability of the linear superposition principle, which in the form presented by him seems to be very complicated and time consuming and from the theoretical point of view at least doubtful.

Comparative calculations of the maximum heave and the maximum subsidence reveal significant differences between the subsidence influence (Pöttgens's based on Grond (1957)) and the heave influence functions.

For example, for a circular extraction area with dimensions $R = H$ and the angle of draw $\gamma = 50$ gon, the maximum subsidence, located directly above the middle of the circular excavation, is equal to the so-called full subsidence (GA = 100%). However, the ground heave calculated for this point is only equal to 29,3% of the maximum heave, obtainable theoretically for an infinitely large area extracted. 90% of the maximum heave is obtained for $R = 10 H$ and 95% for $R = 20 H$.

This means that the graphical integration grid for calculating the heave is much larger than that for calculating subsidence, and the surface heave extends over a much greater area than the mine subsidence. The Grond's method for calculating the relative subsidence over circular mined-out areas (GG) and the Geertsma's method for calculating the relative heave (GG) served Pöttgens to create a nomograph shown in Figure 2.

The nomograph was constructed for the following data:
- Calculating mine subsidence – Grond's method for $\gamma = 50$ gon.
- Calculating heave – $d_m = 0,35 \cdot 10^{-2}$ m^2/MN, $\upsilon = 0,25$, h = 4 M.

Pöttgens's statements and assumptions are in a number of points absolutely logical. His quantitative assumptions are, however, questionable.

One of them is the assumption that the height of the caving zone in the roof is always four times the width of the extraction. According to the results of comprehensive research presented by Borecki and Chudek (1972), the width of the zone falls in the range 3–20 times the mining width. Also the value of the Poisson's ratio is generally accepted as constant at 0.25.

For these reasons the value of the expansion coefficient of $d_m = 0,35 \cdot 10^{-2}$ m^2/MN must be perceived only as applicable to the assumed values of the caving zone height and the Poisson's ratio.

An analysis of the Geertsma solution (1973) shows that for a very large gas deposit the maximum value of surface subsidence is greater than compaction of the deposit itself. The ratio of the two values equals $2 \cdot (1 - \upsilon)$ meaning that for $\upsilon = 0.25$ the maximum subsidence is approximately 50% greater than the compaction of the gas deposit.

In his elaborations on the problem Pöttgens does nor correct this error.

Simulations showed that applying Pöttgens's method to multiple seam mining conditions results, due to extensive differences between the subsidence influence function and the heave influence function, in producing heave distributions substantially different from the calculated subsidence distributions. The discrepancies were seen when comparing such characteristic values as e.g. the maxima.

In the author's opinion the Fenk's method (2000) is useful only in the case of single coal seam extraction and then for estimating the maximum value of ground heave.

The expansion coefficient d_m used in the Fenk's formula can in no way be compared to the symbol expansion coefficient represented by the same in the Pöttgens's equation.

5. MATHEMATICAL MODEL FOR PREDICTING GROUND SURFACE MOVEMENTS CAUSED BY RISING WATER LEVEL IN A FLOODED MINE

The formulae presented below are derived for a finite caving zone element with a base of $\Delta x \cdot \Delta x$ and a height h.

Due to rising of the mine water level the height of the zone increases by Δh. This leads to infinitesimal heave of the ground surface (Figure 3).

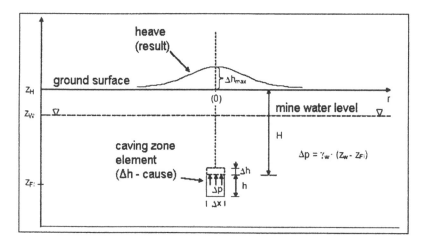

Figure 3. Graphical representations of mathematical equations

Assuming equal cause and result volumes the following solution of Geertsma's influence function is derived:

$$\Delta h(r) = \frac{d_m \cdot \Delta p \cdot \Delta V}{2\pi} \cdot \frac{H}{\left(r^2 + H^2\right)^{3/2}} \tag{11}$$

where $\Delta V = \Delta x^2 \cdot h$; $\Delta p = (z_w - z_{Fl}) \cdot \gamma_w$, and $\Delta h_{max} = \frac{d_m}{2\pi} \cdot \frac{\Delta V}{H^2} \cdot \Delta p$.

Division of the extracted areas into elements allows calculating heave values for any geometry of mining voids by means of linear superposition. Analyses done for multiple seam mining conditions revealed the existence of strong dependence of the expansion coefficient d_m on mining depth and width.

Sroka (1990) and Hejmanowski (1993) applied Knothe's theory to predict subsidence caused by gas and oil extraction. The results of the calculations showed a high level of correlation with the results of in-situ measurements [Sroka 1990]. The subsequent analysis of the parameters of Knothe's theory permited to establish the value of the draw angle β. For Gröningen gas deposit this value equals to approx. 7 gon, from which follows the value of the subsidence influence radius as $R_K = 9H$. Recalculating these values into the values characteristic for a so called Ruhrkohle method results in obtaining the so called radius of full subsidence through equal to 11H, i.e. 11 times greater than the mining depth. The results confirm the validity of the Geertsma solution.

Applying Knothe's theory results in obtaining the formula (12):

$$\Delta h(r) = \frac{d_m \cdot \Delta p \cdot \Delta V}{R_K^2} \cdot \exp\left(-\pi \frac{r^2}{R_K^2}\right) \tag{12}$$

$$R_K = H \cdot \cot\beta,$$

$$\Delta h_{max} = d_m \cdot \frac{\Delta V}{H^2} \cdot \tan^2\beta \cdot \Delta p .$$

Formulae (11) and (12) can be used alternatively to calculate the surface heave for known extraction geometries. Simulations ran for multiple seam mining conditions showed disparity, in some cases substantial, between the obtained predicted subsidence and heave profiles. The presented method allows taking into account numerous geometrical variables of the extracted coal seam dimensions as well as to introduce a mining-depth-dependent expansion coefficient. Further development of the analytical method is determined, however, by extensive research into and in-situ measurements of surface heave caused by mine flooding.

REFERENCES

Baglikow V. 2003: Bergschäden nach Beendigung der Grubenwasserhaltung im tiefen Bergbau. Markscheidewesen 110, 2003, Heft 2, s. 45–49.

Bals R. 1931/32: Beitrag zur Frage der Vorausberechnung bergbaulicher Senkungen. Mitteilungen aus dem Markscheidewesen, 42/43, s. 98–111.

Borecki M., Chudek M. 1972: Mechanika Górotworu. Wydawnictwo Śląsk, Katowice.

Fenk J. 2000: Eine analytische Lösung zur Berechnung von Hebungen der Tagesoberfläche bei Flutung unterirdischer Bergwerksanlagen. Das Markscheidewesen 107, 2000, Heft 2, s. 420–422.

Flaschenträger H. 1938: Die Kostenverteilung bei gemeinsam verursachten Bergschäden im Ruhrgebiet. Mitteilungen aus dem Markscheidewesen, 49, s. 95–137.

Geertsma J. 1973: A Basic Theory of Subsidence Due to Reservoir Compaction: the Homogeneous Case. The Analysis of Surface Subsidence Resulting from Gas Production in the Groningen Area, the Nederlands. Editor Nederlandse Aardolie Maatschappinj B.V., Deel 28, 1973, s. 43–61.

Geertsma J., van Opstal G. 1973: A Numerical Technique for Predicting Subsidence above Compacting Reservoirs, Based on the Nucleus of Strain Concept. The Analysis of Surface Subsidence Resulting from Gas Production in the Groningen Area, the Nederlands. Editor Nederlandse Aardolie Maatschappinj B.V., Deel 28, 1973, s. 63–78.

Goerke-Mallet P. 2000: Untersuchungen zu raumbedeutsamen Entwicklungen im Steinkohlenrevier Ibbenbüren unter besonderer Berücksichtigung der Wechselwirkungen von Bergbau und Hydrogeologie. Praca doktorska RWTH Aachen, Wydawnictwo Mainz.

Goerke-Mallet P., Preuße A. und Coldewey W.G. 2001: Hebungen der Tagesoberfläche über betriebenen und gefluteten Bergwerken. 43. Wissenschaftliche Tagung des Deutschen Markscheider – Vereins, Trier 2001, s. 125–139.

Grond G.J.A. 1957: Ground Movements Due to Mining. Colliery Engineering 34, s. 157–197.

Heitfeld M., Rosner P., Mühlenkamp M. und Sahl H. 2004: Bergschäden im Erkelenzer Steinkohlenrevier. 4. Altbergbaukolloquium, Leoben 2004, s. 281–295.

Hejmanowski R. 1993: Zur Vorausberechnung förderbedingter Bodensenkungen über Erdöl- und Erdgaslagerstätten. Praca doktorska, TU Clausthal, 07.07.1993.

Hejmanowski R., Sroka A. 2000: Time-Space Ground Subsidence Prediction Determined by Volume Extraction from the Rock Mass. Proceedings of the Sixth International Symposium on Land Subsidence, Ravenna, Italia, 24–29 September 2000, Vol. II, s. 367–375.

Mühlenbeck H. 2005: Hebungen nach Einstellung von Wasserhaltungen im Ruhrgebiet. Markscheidewesen 112, 2005, Zeszyt 3, s. 97–102.

Oberste-Brink K. 1940: Die Frage der Hebungen bei Bodenbewegungen infolge des Bergbaus. Glückauf 76, s. 249–256.

Pöttgens J.J.E. 1985: Bodenhebung durch ansteigendes Grubenwasser. 6. Internationaler Kongress für Markscheidewesen, Harrogate 1985, s. 928–938.

Pöttgens J.J.E. 1998: Bodenhebung und Grundwasseranstieg aus geotechnischer und markscheiderisch-geodätischer Sicht im Aachen – Limburger Kohlenrevier. Freiberger Forschungshefte, Bergbau und Geotechnik A 847, s. 193–207.

Sroka A. 1990: Studie zur Analyse und Vorhersage der Bodensenkungen und des Kompaktionsverhaltens des Erdgasfeldes Groningen/Emsmündung. Ekspertyza, Goslar 1990.

Sroka A. und Fenk J. 2003: Studie zu Auswirkungen des Grubenwasseranstiegs auf Bewegungen der Tagesoberfläche in stillgelegten Bergbaubereichen des Saarlandes. F+E – Bericht der TU Bergakademie Freiberg.

Sroka A. 2005: Ein Beitrag zur Vorausberechnung der durch den Grubberwasseranstieg bedingten Hebungen. 5. Altbergbaukolloquium, Clausthal-Zellerfeld 2005, s. 453–464.

Wings R.W.M.G., Misere W.M.H. und Pöttgens J.J.E. 2004: Bodensenkung – Bodenhebung – Bergschäden? 44. Wissenschaftliche Tagung des Deutschen Markscheider – Vereins, Bochum 2004, s. 258–269.

International Mining Forum 2006, Sobczyk & Kicki (eds) © 2006 Taylor & Francis Group, London, ISBN 0415 401178

A Geophysical Model for the Analysis of Seismic Emissions in the Area of Rock Beam Splitting Caused by Mining Operations

Henryk Marcak

AGH – University of Mining and Metallurgy, Institute of Geophysics. Kraków, Poland

ABSTRACT: In Poland the rock-burst risk in underground mines is an important issue in the mining industry. The predictive rock-burst information is extracted from different kind of information based on measurements. One of them is seismic data, but the predictive information included in it is limited, due to small number of strong seismic shocks and the chaotic development of rock destruction before the strong seismic events occur. In the paper, a geomechanical model of rock destruction prior strong seismic shocks is proposed. It can help in an optimal organization of the measurement system and its interpretation for better assessment of the seismic risk in underground mines.

INTRODUCTION

There are two mining regions in Poland subject to rock-burst risk; one is the coal basin in Upper Silesia, the second, the copper region in Lower Silesia. In both regions strong sedimentary layers appear in the roof of exploited parts of the rock masses. They consist of sandstone in the coalmine basin and limestone and dolomite in the copper mining areas.

The significant influence of strong sedimentary layers on building up stresses in exploited rock masses has been considered and accepted many years ago [Saustowicz 1965]. Calculation of elastic deformations, results of rock-beam bending in roof of exploited seam, has been used for modelling stresses in exploited areas. In this paper the stresses produced by rock-beam in the roof of an exploited area are also considered, but its plastic behaviour is analysed.

In the paper it is shown, that both, seismic data, and the results of measurements of changes in ratio of vertical and horizontal deformations of roof layers over the exploited area, can be used in assessment of strong seismic hazard. In the simplest attitude the deformation of roof layer can be considered on the base of elastic theory. The exploited area is here described in a two-dimensional system, the x-axis being perpendicular to the long wall and the z-axis being the depth.

The stress-distribution may thus be described by the equation:

$$\frac{\partial^4 F(x,z)}{\partial x^4} + 2\frac{\partial^4 F(x,z)}{\partial x^2 \partial z^2} + \frac{\partial^4 F(x,z)}{\partial z^4} = 0 \tag{1}$$

where $F_j(x, z)$ is a function, which defines the elements of the stress tensor.

$$\sigma_x(x,z) = \frac{\partial^2 F(x,z)}{\partial z^2} \quad \sigma_z(x,z) = \frac{\partial^2 F(x,z)}{\partial x^2} \tag{2}$$

$$\tau_{xz}(x, z) = \frac{\partial^2 F(x,z)}{\partial z \partial x},$$

and

$$\tau_{i,j} = \begin{Bmatrix} \sigma_x & \tau_{zx} \\ \tau_{xz} & \sigma_z \end{Bmatrix} \tag{3}$$

The components of the displacement vector $\vec{u}$ (u, w) fulfill the relation:

$$2G\frac{\partial u(x,z)}{\partial x} = (1-v)\frac{\partial^2 F(x,z)}{\partial x^2} - v\frac{\partial^2 F(x,z)}{\partial z^2} \tag{4}$$

$$2G\frac{\partial w(x,z)}{\partial z} = (1-v)\frac{\partial^2 F(x,z)}{\partial z^2} - v\frac{\partial^2 F(x,z)}{\partial z^2} \tag{5}$$

where G, v are the stiffness coefficient and Poisson's ratio respectively.

Let the roof layer in the vicinity of the exploited area have a thickness h and elastic coefficients G^1, v1. Harder rocks overlying the seam are treated as a half-plane with elastic coefficients G^2, v^2. This model allows obtaining an elastic model of energy concentration before rock burst. More sophisticated models can be found in literature e.g. [Salomon and Hunro 1967, Bieniawski 1976, Brady and Brown 1993]. The general idea in this approach is to model the stress field and the associated field of deformation by partial differential equations. The equations solved with a finite element method, provide the stress distribution for complex geological structures. There are however, limits in the assessment of real rock deformations by such calculations. Firstly, it is difficult to obtain true initial and boundary conditions. Furthermore, an unstable behaviour of rocks after cracking and fracturing of its masses cannot be properly modelled with equations that assume continuity of the medium.

In polish underground rock bursting mines, recently, a relation between deformation of boreholes located in the roof of the productive layer and emission of strong seismic events has been observed. Sudden jumps of deformations were recorded 5 to 20 days before strong tremors occur. [Orzepowski 1998, Matwiejszyn and Ptak 2000]. Measurement of roof deformation by Orzepowski in coppers mine, with sensors measuring deformations of a borehole diameter in fixed horizontal directions and at fixed depth has been carried out over several years. For example, [Orzepowski 1998] obtained results with sensors installed in four boreholes 30 m above the roof of the ore seam in a copper mine. The results of measurements prior to a strong seismic event are shown in Figure 1. The distance between the shock epicentre and the epicentre azimuth has been estimated and is shown in Figure 1 caption. It can be seen that up to 25.01.97 changes in the borehole diameter were slight. Later deformations of all boreholes exhibited a similar, time varying pattern, until 11.02.97, when a strong mining shock occurred. There is no deterministic relation between anomalous deformations in roof beam and strong seismic occurrence. Only regional deformations (those which are registered with larger number of sensors) are used for prediction. Even, using such criteria only the hazard of strong seismic shocks markedly increases after registration of roof deformations It is significant, that after registration of anomalous deformation, failure of prediction have form of strong events lack after roof deformation not lack of roof deformation before strong seismic shock. For explanation of this results the model of brittle deformation of rocks should be used.

These measurements indicate, that the inelastic deformations have appeared in roof beam. It can be expected, that it have form of plastic horizontal deformations, which can turn into splitting and bending downward of roof layer. The horizontal stresses appear in specific time, when the inelastic deformation appears.

The deformations in strong rocks overlying splitting are the best explained by the use of a model of inelastic deformation. For this purpose Rudnicki's dilatancy-plastic model is suitable [Rudnicki and Rice 1975]. The plastic deformations γ^p and ϵ^p are associated with shear stress τ and compressive pressure p, the elastic deformations γ^e and ϵ^e are linked with the shear stress τ and the compressive pressure p through the coefficients G and K.

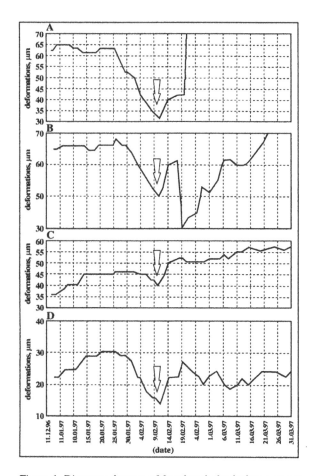

Figure 1. Diameter changes of four boreholes before a strong mining event,
which occurred on 11.02.97. in a copper mine.
a) borehole located 26 m from epicenter, azimuth 83°, b) borehole located 95 m from epicenter, azimuth 15°,
c) borehole located 288 m from epicenter azimuth 78°, d) borehole located 301 m from epicentre azimuth 60°
(from [Orzepowski 1998])

This model assumes that:

$$\gamma^p = \frac{1}{h}(\dot{\tau} - \mu\, \dot{p}) \qquad (6)$$

$$\gamma^e = \frac{1}{G}\dot{\tau} \qquad (7)$$

$$\varepsilon^e = -\frac{1}{K}p \qquad (8)$$

$$\varepsilon^p = \beta\,\gamma^p \qquad (9)$$

$$p = tr\{\sigma_{ij}\} \qquad (10)$$

where $\dot{\tau} = \dfrac{\partial \tau}{\partial t}$, G, K, h are the stiffness, elasticity and plasticity modules respectively, β is the dila-

tancy coefficient, μ is the coefficient of internal friction.

If the energy release is dissipated in the form of heat, and is, therefore, ignored and a loss of cohesion is assumed the following equation may be written.

$$Y\gamma^p + \eta p\gamma^p = \tau\gamma^p - p\epsilon^p \tag{11}$$

where Y is the yield plastic coefficient, η is the coefficient of friction, resulting from the movement of the rock-blocks.

Taking into account the relationship:

$$\epsilon^p = \beta\gamma^p \tag{12}$$

The result of transformation formula (11) and (12) the following formula is obtained:

$$\tau = Y + (\beta + \eta)p \tag{13}$$

and plastic yielding is described by

$$Y = \tau - \mu p \tag{14}$$

where μ is the coefficient of internal friction.

The coefficient η is related to the petrography structure of the rock and the quality of its cementation. The second coefficient is related to "dilatancy". This term means "the increase of the volume of the rock resulting from inelastic deformation". The dilatancy may introduce a negative effect to the μ coefficient. It distorts the relationship between coefficient η and the total stress q^p. The following relationship now emerges:

$$h = \frac{\partial\tau}{\partial g^p} = \frac{\partial Y}{\partial g^p} + \frac{\partial\beta}{\partial\epsilon^p}\cdot p\beta \tag{15}$$

The last equation can be used to describe the bending layer. Movement between the upper and lower planar surfaces of the roof layers is assumed to be a consequence of bending.

$$(\epsilon - \mu\gamma)\cdot(\epsilon - \beta\gamma) + \frac{h}{k}\gamma^2 + \frac{h}{\sigma}\epsilon^2 \le 0$$

The relationship in this beam, based on the dilatancy-plastic model suggested by [Nikitin and Ryżak et al. 1977, 1984] is:

$$(\epsilon - \mu\gamma)\cdot(\epsilon - \beta\gamma) + \frac{h}{k}\gamma^2 + \frac{h}{\sigma}\epsilon^2 \le 0 \tag{16}$$

Applying Legander's method for maximalization left side of equation (16) gives a value for h_o. It can be shown [Nikitin and Ryżak et al. 1997, 1984] that from h_o value the parameters of inclusion zones can be estimated.

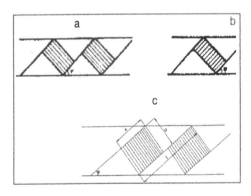

Figure 2. Parameters describing the inclusion zone in the roof rock beam.
a) perpendicular fracturing, b) parallel fracturing, c) parameters of inclusion zone b = x/l

Those zones have the following properties:
- Constrain a number of fractures whose length distribution has fractal properties parallel to inclusion zone boundary or parallel to them.
- They are repeated at regular intervals.

The fracture zones have the following parameters (Fig. 2):

$$(\mu - \beta)^2 = \frac{4h_0}{k} \tag{17}$$

$$\alpha = \frac{\mu - \beta}{2}\left(1 + \frac{h_0}{G}\right) \tag{18}$$

$$\Psi = \frac{1}{2}\alpha\frac{2k}{G + 3k} \tag{19}$$

$$\frac{b}{1} = \frac{G + \frac{3}{4}K}{G + \frac{3}{16}K}\left(1 - \sqrt{1 - \frac{3}{4}\frac{G + \frac{3}{16}K}{G + \frac{3}{4}K}}\right) \tag{20}$$

The direct application of formulas 17–20 to seismic emission in underground mines [Marcak 1985, 1993], proved the relevance of this model for seismic emission in underground mines. The intensity of seismic activity during exploitation has extremes located periodically and the distances between them agree with values described by formulas 17–20.

The inclusion zones are built up by rather complicated fracturing. However, appearance of very strong shocks is rare, and the special mechanism describing their creation should be considered.

FRACTURING IN THE ROOF OF ROCK-BEAM

The rock-burst is a result of rock-mass movements in the fracture surface. The fracturing of the rock overlying exploited seams starts when the difference between the greatest principal stresses σ_I and the smallest one σ_{III} have reached a maximum, called ultimate shear strength. In the case when two dimensional stress distributions describe critical state of stresses, the limit condition has the form:

$$|\tau_{max}| = \frac{\sigma_I - \sigma_{III}}{2} = f(\sigma) \tag{21}$$

The direction of fracture plane can be presented with use of two envelopes in Coulomb-Mohr stress circles as is shown on Figure 3.

Envelopes can be described with equation:

$$|\tau| = \tau_0 + \sigma_\perp \tan\Phi \tag{22}$$

where Φ – angle of internal friction, τ_0 – the cohesive shear strength (depending on the lithological properties of the rock), $\sigma_\perp$ – the stress perpendicular to the fracture plane.

Two points T and T' in Figure 3 indicate two possible planes of fracturing. The angle between them is bisected by the greatest compressive principal stress σ_I. This implies that the angle between σ_I and the plane of fracturing is $\left(\frac{\pi}{4} - \frac{\Phi}{2}\right)$.

According the properties of formula (2) $\sigma_\perp$ can be found geometrically as the σ value of P point where Coulomb-Mohr stress circle touch the limit line while the principal stress can be also found by linking P with the smaller value of Coulomb-Mohr circle on σ axis.

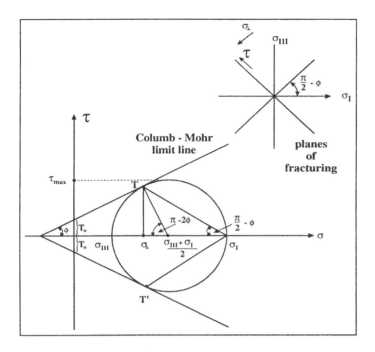

Figure 3. Limit state in Mohr's stress plane

It is also possible to construct a Coulomb-Mohr circle in a $\sigma_\perp$, τ coordinate system on the base of horizontal σ_H and vertical σ_v stresses. The following formulas [Jaeger, Cook 1971] give the relation between the principal stresses σ_I, σ_{III} and horizontal σ_H and vertical σ_v stresses on the roof rock-beam border;

$$\sigma_I = \frac{\sigma_v + \sigma_H}{2} + \sqrt{(\frac{\sigma_v - \sigma_H}{2})^2 + \tau^2{}_{v,H}} \tag{23}$$

$$\sigma_{III} = \frac{\sigma_v + \sigma_H}{2} - \sqrt{(\frac{\sigma_v - \sigma_H}{2})^2 + \tau^2{}_{v,H}} \tag{24}$$

where $\tau_{v,H}$ shear stresses on the horizontal border.

$$\sigma_I - \sigma_{III} = 2 \cdot \sqrt{(\frac{\sigma_v - \sigma_H}{2})^2 + \tau^2{}_{v,H}} \tag{25}$$

In result, taking into account, that $\sigma_I + \sigma_{III}$ is invariant in transformation of stresses, the stresses $\sigma_\perp$, τ can be written in form:

$$\sigma_\perp = \frac{\sigma_v + \sigma_H}{2} + \sqrt{\frac{1}{4}(\sigma_v - \sigma_H)^2 + \tau^2_{v,H}} \cdot \cos 2\Phi \tag{26}$$

$$\tau = \sqrt{\frac{1}{4}(\sigma_v - \sigma_H)^2 + \tau^2_{v,H}} \cdot \sin 2\Phi \tag{27}$$

The critical state in terms of σ_v, σ_H for fixed $\tau_{v,H}$ can be reached on two ways (Fig. 4). In one solution, called the active state (described in Figure 4 with sign (1)) σ_H is decreasing to the smallest value (the rock is under extension). In other solution σ_H increase up to the highest value $\sigma_H^{(2)}$ (the rock is under compression) as is shown with the use of a large circle.

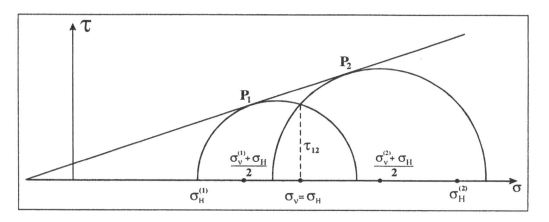

Figure 4. Two ways of reaching the limit condition

The construction shown on Figure 5 gives a method for obtaining principal σ_I direction, when stresses σ_v, σ_H, $\tau_{v,H}$ are acting a horizontal planer element. The directions of principal stresses for active and passive solutions are shown in Figure 6.

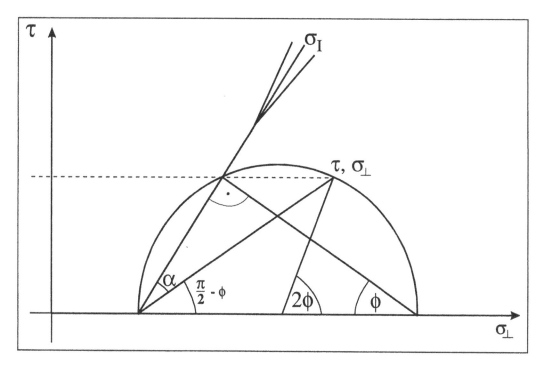

Figure 5. Construction for finding principal stress direction σ_I for the planer element represented by a normal and shear stress, τ and $\sigma_\perp$

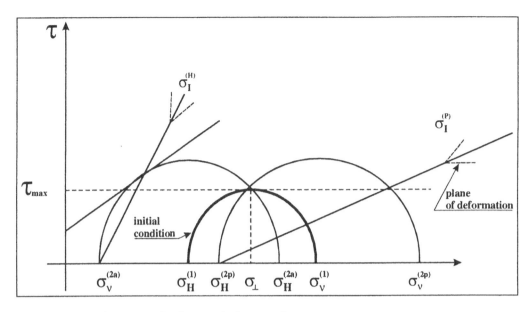

Figure 6. Two modes of stress development in the strong layer
over the roof in sedimentary overlying rocks.
1 – passive state, 2 – active state

It seems to be important, that the principal stresses are turning from horizontal into a markedly more vertical direction, if the situation is changing from passive into active. The directions of fracturing surfaces are also rotated and are assuming vertical directions. Also in the active state the value of τ needed for reaching the limit line is smaller then for the passive state. Coming back to the situation, in Polish underground mines, the bending of roof layers causes elastic/plastic [Ramsey, Huber 1987] behaviour at weak lithological borders. That deformation has elastic and plastic stages. After extending the plastic limit the limit shear line in the Coulomb-Mohr circle in plastic material is parallel to the horizontal axes.

If the deformation takes places in two layers one fractured, weak, lower in the roof of exploited volume and second strong, homogeneous, overlying the first the following development of deformations can be expected. The empty volume, which is left after exploitation becomes unfilled with first layer, as a result of rock subsidence. The first layer is squeezed out due to the shearing action of the plastic underflow of the roof layer. The strength of the roof material in this situation is virtually independent of the effective confining stress; the limit line of this material therefore is parallel to the σ axes. The underflow of the roof layer (which can be recognized by measurements of the roof deformations as it was shown in Figure 2) actively influences the stresses in the overlying layer. The limit condition may be obtained in two different ways. If drag forces in the upper, strong layer are produced as a result of overflowing by the strength of the rock overlaying the lithological border, they should increase, with an increase in the underflow layer extension (in the case of excavation, the area of open roof over the excavated seam). Then the limit state is reached for very large values of loading stresses and due to the almost horizontal position of $\sigma_I^{(a)}$, (passive state the planes of fracture are also horizontal.

Such development of deformation is rather rare. Splitting on the lithological border can cause release the horizontal stresses and the risk of turning from a passive to an active state and the upward turning of the split is large. If weak zones such as faults or lines of fault extension, zones of mechanical changes, edge in the overlying rock mass, or axe of synclines are located in places, where the plastic deformation along the strong border layer is bending upward, in result of changes the stress state, the creation or activation of a seismic zone can be expected. The zones, if activated, can develop through migration into an elongated strip. This process is shown in Figure 7. The activation of seismic zones is often observed in copper mines in Poland. An example of shock distributions (one from many registered) in copper mine is shown in Figure 8.

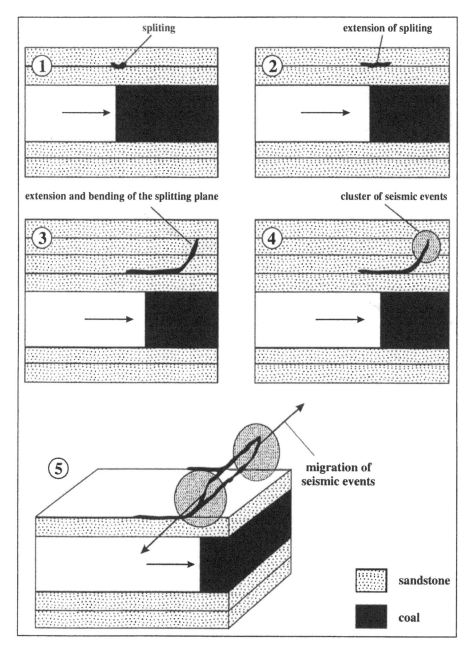

Figure 7. The geo-mechanical model, showing the development of rock destruction, which leads to creation of elongated zones of seismic emission

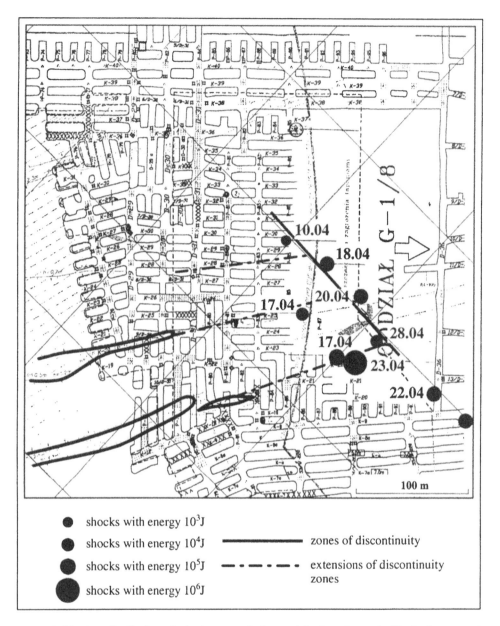

Figure 8. The time distribution of seismic events during exploitation of stope in "Rudna" copper mine. Continuous line shows the number of all shocks recorded during 10 days. Appearance of shocks with energy 10^4 J, 10^5 J and 10^6 J are shown with the circles lines

Shocks are located along the linear zones that follow existing faults or their extension. It is also possible, that in spite of roof deformation the limit condition is not reached. In that situation there is not seismic emission after anomalous deformation of roof layer.

Hitharto the discussion of rock-mass deformation prior to strong seismic shocks have been directed into the location of strong shocks. Also the time aspects of such events should be taken into consideration.

ACTIVATION OF SEISMIC ZONES

In mining conditions the process of seismic energy release is demonstrated by the slippage of one rock block against another. Slip zones appear to be result of inelastic deformations. In such zones [Rice and Ruina 1983, Rice 1980] the shear stress τ in the zone increases very rapidly (Fig. 9) in the first stage of deformation called "hardening" and a slight increase of the shear deformation along the zone of failure is observed. If τ_p denotes shear the peak slip resistance τ_f residual frictional shear resistance, which results after a suitably large slip. The relation between the slip value and τ have relation shown in Figure 9.

Shear resistance can start from:

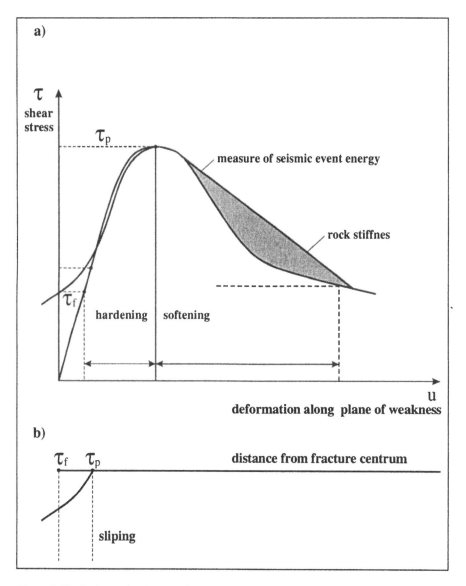

Figure 9. Hardening and softening of rock before seismic emission (modified from [Rice 1980])

In this stage the ultimate shear stress τ_p is reached. In next stage called "material softening", τ intensively decrease and the ratio of τ decreasing determines the amount of energy, which is released in the seismic event (the energy is a measure of difference between the τ stress decrease and the stiffness of material shown in Figure 9 (as a straight line over shadowed area) and depends on the difference between τ_p and τ_s. As it is shown in Figure 9 the ratio of slip depends on the stage of fracturing and the distance from slipping surface. Increase of shear resistance τ in hardening stage is associated with markedly smaller deformation then the deformation expected in result of residual shear resistance. It means, that before seismic energy is released the slipping ratio should decrease in the region of the seismic zone in comparison with the slipping in other regions. Also the slipping ratio should depend on time and lessen just before rock-burst.

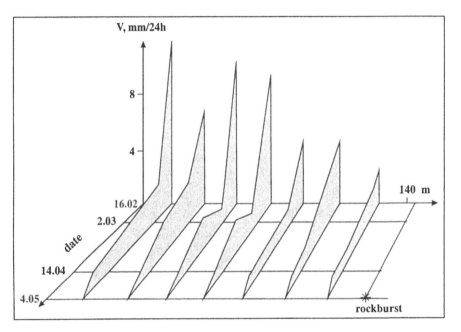

Figure 10. The velocity of rock roof subsidence before rock-burst in "Szobmbierki" coal mine [Wanior 1984]

Figure 10 shows the rate of subsidence before strong rock bursts [Wanior 1984]. Measurements of the roof deformation ratio were taken at regular intervals (x axis) along a roadway at different times (shown on the y axis). The ratio markedly decreases near to the shock epicentre location, and just before the rock-burst occurrence. Confirming the existing of a hardening stage in the deformation of the rock-mass before the rock-burst. This behaviour confirms that model shown in Figure 9 is also valid in the case of mining rock-bursts. From this consideration it can be concluded that seismic energy depends on the ultimate shear stress and the rate of the shear stress decrease after reaching τ_p shear stress.

The deformation within fracture zones is different from that in neighbouring areas. Asperities and rock bridges prevent radical movements of the rock block. The process of destroying these obstacles and the extension of the zone is related to the associated intensive seismic emissions. The slippage ratio can be related to the seismic activity (which is the number of seismic events registered in unit time). After intensive creeping in the beginning of preparation for a strong seismic event the slipping ratio decrease should be correlated with a dropping in seismic activity.

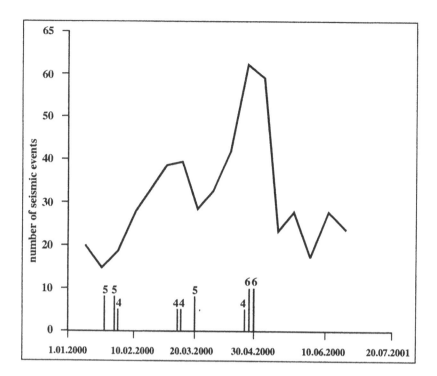

Figure 11. The distribution of strong seismic shocks (with energy over 10^5 J) in "Rudna" copper mine stope

Figure 11 shows the time distribution of shock emission activity from a copper mine stope. The seismic activity (number of shocks in 10 days) has increased before the appearance of strong seismic events and decreases during the actual occurrence of strong seismic events. For example before the strong seismic events with the highest energy dated 30.04 the activity increased from 30 to 65 and then dropped to 30 at the time of the strong seismic events.

CONCLUSIONS

There is a large level of uncertainty in the predictive information intrinsic in measured data, which is observed before strong seismic shocks. In Polish underground mines, where strong rock layers appear in the roof of exploited parts of the rock masses, the level of uncertainty can be reduced if the prediction of a strong seismic event is based on space and time models of rock deformations preceding it.

The following sequence of observations should be planed in areas where strong seismic shocks are expected:
– Anomalous deformations in roof layers of the exploited area.
– Seismic emissions along a linear seismic zone.
– Sudden fluctuation of seismic emission in seismic zone.
– Substantial declining in roof subsidence at the time and place where the seismic shock is likely to appear.
– Decrease of seismic activity before strong seismic events occur.

All these factors, if interpreted together, can reduce uncertainty in prediction of rock-burst in mines.

REFERENCES

Bieniawski Z.T. 1973: The Determining Rock Mass Deformability Experiences for Case Histories. Inst. J. Rock Mech. Min. Sci. and Abst. 14, pp. 237–242.

Brady G.H.G. and Brown E.T. 1993: Rock Mechanics for Underground Mining. Chapman and Hall, New York.

Jaeger J.C., Cook N.G.W. 1971: Fundamentals of Rock Mechanics. Chapman and Hall LTD London.

Marcak H. 1985: The Geophysical Models of the Rock Destruction Process Development Prior to Rock-Burst and Shocks in Underground Mines. Publ. Inst. Geophys. Polish Acad. of Science, M-6, pp. 149–173 (in Polish).

Marcak H. 1993: The Use of Pattern Recognition Method for Prediction of the Rock-Burst. Proc. of the III Int. Symp. on Rock-Bursts and Seismic in Mines A.A Balkema, Rotterdam, pp. 223–226.

Matwiejszyn A. and Ptak M. 2002: Measurement of Borehole Deformations for Assessment of Rock-Burst Hazard in Seismogenic Process Monitoring. Ogasawara H., Yanagidani T. and Ando M. eds, Balkema, Rotterdam (in print).

Nikitin L.W., Ryżak E.T. 1984: An Model of Forming the Systems of Tectonic Fractures. Geodinam. Tssled 7, pp. 56–76 (in Russian).

Nikitin L.W., Ryżak E.T. 1987: The Rules for Rock Destruction Based on the Internal Friction and Dilatancy. Fizyka Ziemi nr 5, pp. 22–38 (in Russian).

Orzepowski S. 1998: The Attempts to Estimate the Rock-Burst Hazard on the Base of Deformation the Vertical Boreholes and its Relation to Recorded Seismic Activity. In.: Proc. of the XXI Winter School of Geomechanics. AGH – University of Mining and Metallurgy, Krakow, pp. 349–367 (in Polish).

Ramsay J.G., Huber M.I. 1987: The Techniques of Modern Structural Geology. Academic Press, p. 700.

Rice J.R., Ruina A.L. 1983: Stability of Steady Frictional Slipping Trans. ASME, J Appl. Mech. 50, pp. 343-349.

Rice J.R. 1980: The Mechanics of Earthquake Rupture in Physics of the Earth Interior. Proceedings of the International School of Physics "Enrico Fermi", Italian Physical Society, North-Holland, Pub. Co, pp. 515–649.

Salomon W.D.G. and Munro A.H. 1967: Study of Strength of Coal Pillars. SAIMM. 68, pp. 55–67.

Sato K. and Fujii Y. 1988: Induced Seismicity Associated with Longwall Coal Mining. Inst. J. Rock Mech. Min. Sci. and Geomech. Abst. 25, pp. 237–242.

Saustowicz A. 1965: Principals of Geomechanics. Wyd. Śląsk, Katowice (in Polish).

Wanior J. 1984): The Method of Prediction of Rock-Bursts and Mining Tremors on the Base of Geodetic Measurements. Proceedings of the Conference PTPNoZ, Częstochowa (in Polish).

International Mining Forum 2006, Sobczyk & Kicki (eds) © 2006 Taylor & Francis Group, London, ISBN 0415 401178

Monitoring of Natural Hazards in the Underground Hard Coal Mines

Władysław Mironowicz, Stanisław Wasilewski
Research and Development Centre for Electrical Engineering and Automation in Mining EMAG. Katowice, Poland

ABSTRACT: The natural hazards especially methane, fire and rock-bump hazards make the most serious danger for present-day mines and have a crucial effect on miners' safety and continuity of mining operations. Degree of these hazards increases with concentration of coal faces and use of high-duty mining methods for seams lying deeper and deeper. The interaction of natural hazards at seams liable to rock-bumps may lead to intensity both fire and spontaneous methane emission. The methane continuous monitoring systems in the range of $0 \div 100\%$ CH_4 including automatic power-off as well as the early fire detection systems based on measurements of CO, CO_2, O_2 and smoke are nowadays a standard. The improvement in efficiency of mine rescue operations can be achieved by quick reaction to the hazards. The issue concerns e.g. the mines in which the associated natural hazards may occur and where the bumps of definite energy requires an immediate power-off not only in the hazardous areas but also at ways of air flow and propagation of methane disturbances. The up-to-date rock-bump hazards monitoring systems use seismoacoustic and micro-seismologic methods characterized by high dynamics of signals to be registered. Work safety of miners in the underground areas means also the systems of their localization and attendance in mine workings as well as the warning systems in case of hazards e.g. fire or gas and smoke propagation. The miners' localization systems and load-speaking and alarm broadcasting communication systems should be therefore disseminated to support a mine operator to remove the staff from hazardous areas. Detection of the state of emergency requiring mine rescue operations allows a mine operator to activate the underground signalling devices and banners showing safe escape routes.

INTRODUCTION

The natural hazards have a crucial impact upon the safety of the mines' crew and continuity of routing mining operations at the mines. Degree of these hazards increases with extraction of seams at ever greater and greater depth use of heavy-duty extraction technologies. Monitoring of technological process at mines is essential from the viewpoint of supervision and control of the process's course. Its superior objective consists in securing of continuity of the machines operation and proper operation of technological links starting from a coal winning machine and ending at coal sale and loading points at coal preparation and washing plants on the surface. It is obvious than monitoring regards not only the machines directly connected wit the production but also power supply, electrical protection and media. Monitoring and control focus on enlargement of work time and efficient diagnostics of machines utilising signalling and visual presentation of the process's parameters.

Still more important is the control and monitoring of safety conditions, especially in underground mines. At present conditions running of safe exploitation requires complex solutions with utilisation of modern and reliable systems of monitoring, control and automatic safeguarding. In this case the utmost importance objective consists in securing of safe work conditions for the miners engaged in underground workings. That regards both the current and random monitoring of parameters and assisting the activities of the mine's dispatchers and supervision during rescue works connected with saving of miners and protection-prevention activities.

Monitoring and control of natural hazards were treated up-to-date as autonomous subsystems of control and safeguarding. The experiences of recent years, unfortunately those tragic ones showed that in order to secure efficient methane-fire hazards prevention activities it is necessary to functionally combine all in one system. Taking the account of spatial structure of quakes origination or source of coal spontaneous heating and also occurrence of methane sources becomes indispensable. The effectiveness of safety status monitoring and warning the crew about occurring hazards as well as their rescue in disastrous situations require the engagement of the most modern information and microelectronic techniques as well as audio-visual means in modern systems of methane-fire natural hazards monitoring. On the other hand it is necessary to integrate the systems of methane – fire natural hazards monitoring, monitoring of geophysical or water hazards with the alarm loudspeaking/signalling systems and the system of locating the miners' whereabouts.

1. MONITORING AND CONTROL OF GAS HAZARDS

Solutions with regard to methanometric safeguarding have a decisive importance in the assessment of functionality and effectiveness of monitoring and control systems of gas hazard. The systems of automatic methanometry decide on the one hand about the safety but on the other hand have also considerable influence upon the continuity of operation of coal winning machines by switching off of electric energy in hazardous zones.

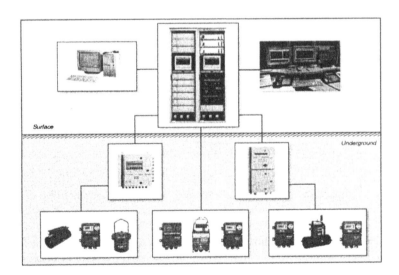

Figure 1. Pictorial diagram of SMP-NT system

Irrespective of reduction of the total number of reported fires at Polish hard coal mines the fire hazard does not decrease. That specially regards seams prone to rock bumps where both natural quakes and distressing of the strata due to mining operations as well as due to distressing shotfireing done during hazard prevention activities increase the susceptibility to spontaneous combustion. Despite of this, thanks to, among others, application in recent years of automatic CO-metry it is possible to early detect a fire hazard and suppress it at its initial stage of development without the necessity of announcing of an emergency firefighting operation.

After a disaster at "Halemba" mine in 1991 (during which due to methane explosion 19 miners lost their lives) the investigations indicated that in the areas with high dynamic of methane emission new solutions of methanometric safeguarding are necessary. That require substitution of the old generation systems, with 4-minute sampling time commonly applied at Polish mines, with new solutions of continuous measurement mode and allowing the measurement of methane in the range of 0÷100% CH_4 and faster switching off of electric energy in the hazardous zone. The idea of automatic system of ventilation control is based on continuous monitoring of the mine air parameters. The system applies a number of air parameters sensors, including, among others: CH_4, CO, CO_2, smoke, air velocity, absolute pressure and pressure drop, air and the strata temperature and also dust sensors. Basing on this idea the system of monitoring and control of methane-fire hazard of SMP-type was elaborated by EMAG Centre (Fig. 1) and put into use in 32 mines.

1.1. The SMP methane and fire monitoring system

This state-of-the-art dispatcher system is a solution that utilises Polish expertise in the field of safety and control of gas and fire hazards in the underground coal mines. The system realises continuous monitoring of significant parameters of the air, and is of modular structure that enables to adapt its functionality to the needs and size of a mine being protected. The solutions applied here are on a par with the most advanced ones in the mining industry throughout the world. All elements of underground equipment are intrinsically-safe and may operate in explosive atmosphere, and central power supply of underground devices from the surface facilitates the system's uninterrupted operation even in the event of electric power supply breakdowns (also at the time of disasters) or failures in a mine underground.

The idea of automated system for ventilation control is based on continuous monitoring of the mine air parameters. For this purpose, in a mine ventilation network the air parameters sensors that provide current information on any variations of the mine air composition are installed.

In this system a number of the air parameters sensors, are used, among them:
- Methane-monitors.
- Carbon monoxide analysers.
- Smoke detectors.
- Stationary oxygen-meters.
- Stationary anemometers.
- Air temperature sensors.

The system configuration shall depend on specific hazards pertaining to the particular mine and positioning of sensors in mine workings results directly from the mining industry regulations and the particular mine requirements.

The CCD-1 microprocessor local underground stations operate as data concentrators. The underground station intermediates in data transmission between the underground sensors and surface station, and provides remote supply of sensors. The CCD-1 underground station is powered from the surface. The station enables to measure continuously-the mine atmosphere parameters, to monitor binary signals and also to control equipment. The station has 8 analogous inputs, 16 binary inputs and 4 control outputs.

The surface central station that realises power supply of underground equipment and data transmission consists of a block of data transmission systems and supply, and a master computer system. Central computer of the SMP system is a local dispatch station (Fig. 2) for the mine safety status.

Figure 2. Methane and fire hazard monitoring system type SMP –Central Station

The SMP dispatch system, implemented successfully in the hard coal mines, guarantees, among others:
–   Continuous monitoring and control of methane and fire hazards as well as other ventilation parameters.
–   Central power supply of all underground elements from the surface.
–   Automatic switching-off of the electric power at mine workings in case of emergency and breakdown.
–   Early detection of fires.
–   Supporting a dispatcher during rescue operation after alarm announcement, and supporting ventilation service at the time of fire extinguishing and prophylactic actions.
–   Monitoring and control of parameters of main fans and energy consumption.

This modern mine control system is a solution completely utilising Polish experiences with regard to safety and monitoring of gas and hazards in underground mines. The system realizes continuous measure of essential air parameters and possesses a modular structure allowing for its adjustment to the size of the mine to be protected. The accepted solutions belong to the most modern in the scale of the World mining. All underground devices and instruments are intrinsically safe with respective certificates and can operate in explosive atmosphere, and the central powering of underground devices and equipment from the surface allows for non-interrupted operation of the system, even in the cases of switching off of electric energy (also during disasters) or failure in electric power supply in the underground of the mine.

2. SYSTEMS OF SEISMIC HAZARDS MONITORING

The effectiveness of natural hazards fighting requires a development of new method of hazards assessment prevention activities and prediction of their random occurrences. New methods of rock bump hazard monitoring require today the introduction of methods based in registration of dynamic phenomena and static ones taking place deep in the roof, registered in long diameter boreholes and also analyses of the borehole's deformation in a spatial layout as well as the analysis of signals of great dynamics and brad spectrum of frequency.

For many years the EMAG Centre elaborates and develops systems for registration of crashes and quakes with the use of geophones in seismoacoustic systems (SAK, ARES – Fig. 3) and seismometers in the systems of micro seismology (SYLOK, ARAMIS – Fig. 4). Majority of Polish hard coal and copper mines apply some of these solutions. Development in this sphere regards technology in which mainly development of probes' parameters, systems of transmission and methods and algorithms of processing. The most modern solution included digital seismoacoustic system of ARES-4D type and digital microseismic system of ARAMIS-M type with tri-axial probes utilising digital transmission of the date type DTSS.

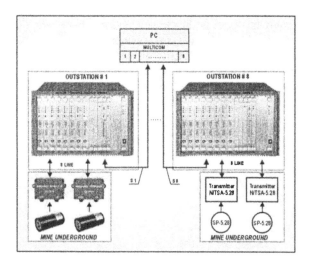

Figure 3. Seismic-acoustic system type ARES

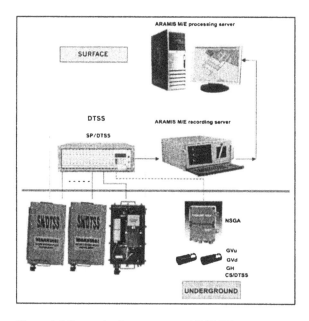

Figure 4. Micro-seismic system type ARAMIS

Improvement of the effectiveness of seismic systems with regard to monitoring of quakes and location of their sources will be also possible after integration as early as at the stage of interpretation and processing of signals from seismoacoustic and seismological systems, as well as by presentation of the registered events on a uniform spatial diagram (Fig. 5).

3. INTEGRATION OF SAFETY SYSTEMS

Safety of work for miners underground also means the system of warning them in case of occurring hazards, for instance, in case of a fire or propagating admixtures of gases also after a bump. Such systems need the propagation of the systems for locating the whereabouts of the miners, the system of loudspeaking communication and the alarm signalling system which, in case of occurrence of a hazardous situation, will be assisting the mine's dispatcher in withdrawal of the crew from hazardous zones of the mine. Effectiveness improvement of endangered miners rescue system is possible due to a rapid reaction to a hazard that has just occurred and immediate transfer of messages in writing or evacuation signals.

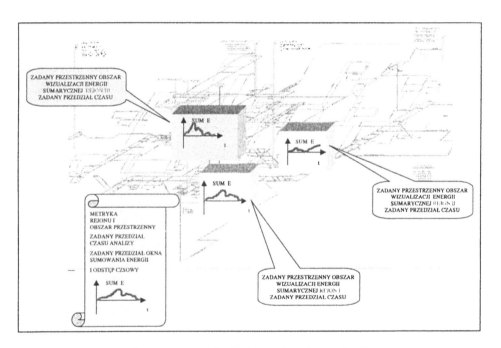

Figure 5. Visualisation of tremors energy distribution on the mine spatial diagram

Such functions are being realised by a completely integrated system of the STAR-SMPZ type introduced for the first time at "Bogdanka" mine (Fig. 6).

In the case of mines with interacting hazards the occurrence of a bump of certain energy requires immediate switching off of electric energy not only in the affected zone but also along the route of airflow towards the uptake shaft. Therefore the structural integration of the safeguarded areas on the base of a spatial diagram of the mine is necessary and also the integration of the safeguarding systems, by now the autonomous ones, into one system (bump + methane). Such a solution is being realised at present at "Bielszowice" mine (Fig. 7).

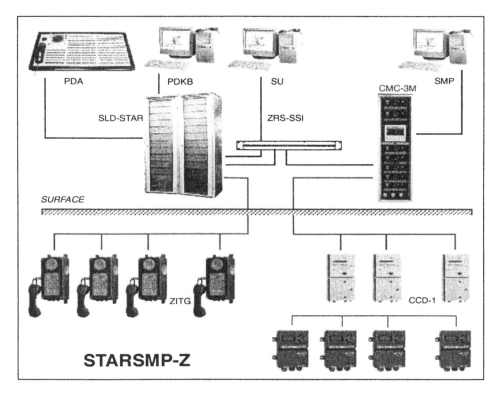

Figure 6. Integrated system of monitoring and control of safety and alarm-loudspeaking/signalling

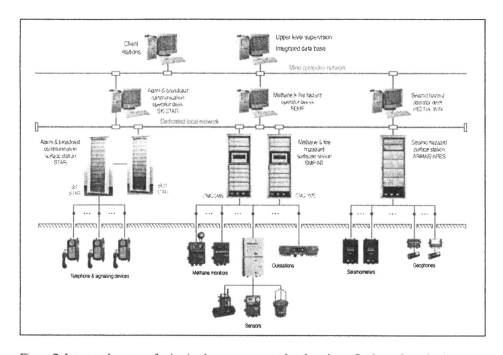

Figure 7. Integrated system of seismic phenomena control and methane-fire hazard monitoring

4. CONCLUSION

The above-presented contemporary mine control system applies the most advanced development of information technology and visualisation. The natural hazard monitoring system proposes the innovative conception of mine management, based on distributed structure of specialised sub-systems (methane & fire or geophysical) dedicated to the particular mine services. The high degree of hazards calls for a new approach to monitoring and control of ventilation. The test results of the methane signals point out to the necessity of using continuous monitoring equipment in mines.

The expert system and new software tools allow for complex interaction of ventilation, staffs rescue operations, fire-fighting and prevention. Improvement of the effectiveness of seismic systems with regard to monitoring of quakes and location of their sources will be possible after integration seismoacoustic and seismological systems and computer interpretation and processing of data.

Effectiveness of control of safety state and staff warning on occurring hazards as well as rescue operation in the event of disasters requires integration of the natural hazards, monitoring systems with systems for warning/broadcasting and localisation of mine's staff. The prototype system (of features described in this paper) with a synoptic projection table, implemented in the hard coal mine "Bogdanka", indicates the new direction for development of dispatch systems for process supervision and control of a mine safety state.

REFERENCES

Isakow Z. 1999: New Forms of Visualisation in Sispatch Systems for Bump Hazard Assessment. Proceedings of Science & Technology Conference on: Dispatch Systems for Monitoring Technological Processes and Safety in Mining Industry Sector. Mechanisation and Automation in Mining, No 4–5/344.

Krzystanek Z., Mirek G., Wojtas P. 2000: Integration of Alarm-Broadcasting System with Methane-Fire Monitoring One as Method for Improving Mine Safety. Mechanisation and Automation in Mining, No 9–10/358.

Mironowicz W. 2000: Advanced System for Process Monitoring for the Hard Coal Mine "Bogdanka". Proceedings of Conference 25th Anniversary of Lubelskie Zagłębie Węglowe (Lublin Hard-Coal Basin), Lublin.

Wasilewski S. 1997: Natural Hazards in Polish Hard Coal in the Light of Disasters and Accidents in the Period of 1960–1994, Part 3. Monitoring and Control of Hazards. Work Safety and Environmental Protection in Mining, WUG, No 3 (31).

Wojtas P., Mirek G. 1999: Visualisation of States and Processes Integration on the Example of Dispatch Systems STAR and SMP. Proceedings of Science & Technology Conference on: Dispatch Systems for Monitoring Technological Processes and Safety in Mining Industry Sector. Mechanisation and Automation in Mining, No 4–5/344.

International Mining Forum 2006, Sobczyk & Kicki (eds) © 2006 Taylor & Francis Group, London, ISBN 0415 401178

Multiple Data Collection System in CO_2 Injection Investigation in Upper Silesian Deep Coal Seams

Paweł Krzystolik, Jacek Skiba, Bartłomiej Jura
Central Mining Institute of Katowice. Poland

ABSTRACT: The parameters, being subject of measurement during the CO2 sequestration into the Upper Silesian coal seams under the RECOPOL project can be divided into three groups: technological, safety and environmental-ones. The technological parameters mainly refer to the constant monitoring of the pressures inside the CO_2 storage tanks, at the inlet and outlet of the pump skid, at the injection well wellhead, periodically inside the well (at different depths – by the pressure gauges). Apart from above also temperature at the wellhead and the flow of CO_2 are being constantly measured. As far as safety parameters are concerned they can be divided into two groups again: the surface and the underground-ones. The underground parameters are very important as they refer to the constant monitoring of CO_2 content in underground workings of the neighbourhood coal mine "Silesia". The surface safety parameters measurement is focused on periodical measurement of MS-4 well gas content, its concentration and flow, measurement of water level in the a.m. well, its PH and conductivity.

The last but also of great significance is the measurement of environmental parameters. In RECOPOL project case this is constant measurement of CO_2 content in the five surface wells (at the depth of 2 meters from the ground surface). These measurements aim is to provide us with the information whether there is any breakthrough of CO_2 from the injected coal seams to the ground surface. Only complex analysis of all above-mentioned parameters can give full understanding of CO2 sequestration phenomena and can drive through the scale of the problems, which have to be faced.

INTRODUCTION

According to many reliable literature sources of information, the Earth's surface temperature has risen by about 0.6 degree Celsius in the past century, with accelerated warming during the past two decades. There is new and stronger evidence that most of the warming over the last 50 years is attributable to human activities. Human activities have altered the chemical composition of the atmosphere through the build-up of greenhouse gases – primarily carbon dioxide, methane, and nitrous oxide. The heat-trapping property of these gases is undisputed although uncertainties exactly how earth's climate responds to them.

Energy from the sun drives the earth's weather and climate, and heats the earth's surface; in turn, the earth radiates energy back into space. Atmospheric greenhouse gases (water vapour, carbon dioxide, and other gases) trap some of the outgoing energy, retaining heat somewhat like the glass panels of a greenhouse. Without this natural "greenhouse effect", temperatures would be much lower than they are now, and life as known today would not be possible. Instead, thanks to greenhouse gases, the earth's average temperature is a more hospitable 15.6°C. However, problems may arise when the atmospheric concentration of greenhouse gases increases. Since the beginning of the industrial revolution, atmospheric concentrations of carbon dioxide have increased nearly 30%, methane concentrations have more than doubled, and nitrous oxide concentrations have risen by about 15%. These increases have enhanced the heat-trapping capability of the earth's atmosphere. Why are greenhouse gas concentrations increasing? Scientists generally believe, that the combustion of fossil fuels and other human activities are the primary reason for the increased concentration of carbon dioxide. Increasing concentrations of greenhouse gases are likely to accelerate the rate of climate change. Scientists expect that the average global surface temperature could rise 0.6––2.5°C in the next fifty years, and 1.4–5.8°C in the next century, with significant regional variation. Ocean evaporation will increase as the climate warms, which will increase average global precipitation. Soil moisture is likely to decline in many regions, and intense rainstorms are likely to become more frequent. Sea level is likely to rise 0.6 meter along most of the coasts.

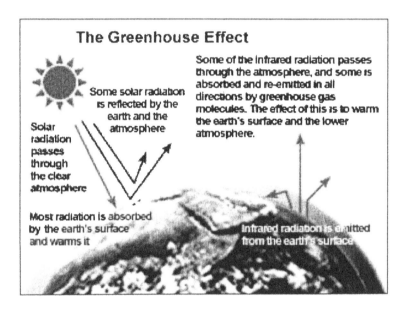

Figure 1. Phenomena of Greenhouse Effect

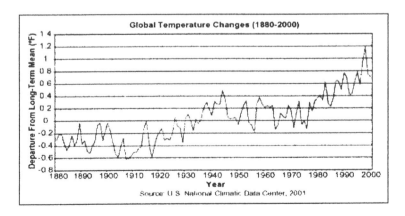

Figure 2. Global temperature changes (1880–2000)

GLOBAL ENVIRONMENTAL POLICY ISSUES

Worldwide, there is increasing concern regarding climate change issues related to Green House Gas (GHG) emissions. It is also recognized, that global issues require global responses. This app-roach has bean reflected in the Kyoto Protocol, with many major nations agreeing to limit green-house gas emissions in the period up to 2012. Thus the European Union intends to decrease its GHG emissions by 8% in 2008–2012 compared to 1990 under the Kyoto Protocol. This will be achieved through a burden sharing agreement between Member States taking into account the fuel mix and situation pertaining within each State. The EU is making reasonable progress so far, al-though there is still uncertainty on whether a reduction by 8% will be achieved in the designated timescale.

CO₂ STORING

Storing industrially generated CO2 in deep underground formations is being seriously considered as a method for reducing greenhouse gas emissions to the atmosphere. Growing interest has lead to significant investment by governments and the private sector to develop this technology and to evaluate if this approach to greenhouse gas control could be implemented safely and effectively. Depleted oil and gas reservoirs, coal beds and deep brine-filled formation are all being considered as potential storage options. Depleted oil can gas reservoirs are particularly suitable for this pur-pose as they have been shown by the test of time that they can effectively store buoyant fluids, such as oil, gas and CO_2. Storage in deep brine-filled formations is in principle the same as storage in oil or gas reservoirs, but the geologic seals that would keep the CO_2 from rising rapidly to the ground surface need to be characterized and demonstrated to be suitable for long-term storage. Coal beds offer the potential for a different type of storage, where CO_2 becomes chemically bound to the solid coal matrix. Over hundreds to thousands of years, some fraction, including possibly all of the CO_2, is expected to dissolve in the native formation fluids. Some of the dissolved CO_2 would react and in the future become part of the solid mineral matrix. Once dissolved or reacted to form minerals, CO_2 is no longer buoyant and consequently, would no longer rise rapidly to the ground surface in the absence of a suitable geologic seal. As illustrated in Figure 3, these two fundamental ideas are the basis for secure geologic storage of CO_2 and the broad context into which the role of, and technologies for, monitoring must be assessed.

Reliable and cost-effective monitoring will be an important part of making geologic sequestra-tion a safe, effective and acceptable method for greenhouse gas control. Monitoring is likely to be required as part of the permitting process for underground injection and will be used for a number of purposes, namely, tracking the location of the plume of injected carbon dioxide, ensuring that injection and abandoned wells are not leaking, and for verification of the quantity of carbon dioxi-de that has been injected underground. Additionally, depending on site-specific considerations, monitoring may also be required to ensure that natural resources such as groundwater and eco-systems are protected and that local populations are not exposed to unsafe concentrations of carbon dioxide. There are many methods that are available for monitoring carbon dioxide in surface and subsurface environments for on-shore geologic storage sites. Methods for monitoring the subsurfa-ce environments include geophysical techniques such as the time-lapse 3-D seismic imaging that has been used successfully at Sleipner[1] and the high-resolution cross-well seismic imaging that has been used to monitor carbon dioxide behaviour in EOR projects. In addition, the potential for other geophysical methods such as electromagnetic imaging, gravity and tilt-meters are discussed. For monitoring geochemical interactions between carbon dioxide and the geologic formation, natural and introduced tracers, major ion geochemical indicators and pH are discussed. Methods for mo-nitoring carbon dioxide concentrations and fluxes on the surface range from conventional flow me-

97

ters and simple carbon dioxide sensors, to the potential for future applications of remote sensing and laser-based techniques for detecting carbon dioxide dispersed in the environment. Measurements intended to detect seepage back into the atmosphere or to detect ecological impacts would be quite different for different environments, which are meant to be used as a potential place of injection and need to be addressed separately.

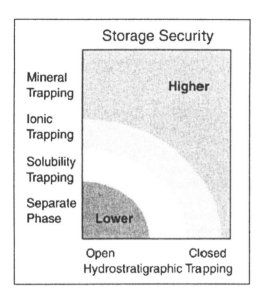

Figure 3. Physical and geochemical processes that enhances storage security

INTRODUCTION INTO RECOPOL PROJECT

The RECOPOL project was an EC-funded research and demonstration project to investigate the technical and economic feasibility of storing CO2 permanently in subsurface coal seams. An international consortium was formed to execute the research, design, construction and operation of the RECOPOL project. This consortium is formed by research institutes, universities and companies from the Netherlands, Poland, Germany, France, Australia, USA and by the IEA Greenhouse Gas R&D Program. Feedback to the project from industry and governmental organizations is realized through the participation of Shell International, JCoal (Japan) and the Federal Region of Wallonie (Belgium) in an end-user group. Overall co-ordination of the project is in the hands of TNO-NITG. The main aim is to demonstrate, that CO_2 injection in coal under European conditions is feasible and that CO_2 storage is a safe and permanent solution before it can be applied on a larger scale in a socially acceptable way. This is the first field demonstration experiment of its kind in Europe. The development of the pilot site in the Upper Silesian Basin in Poland began in summer 2003. One of the existing coalbed methane wells was cleaned up, repaired and put back into production. As RECOPOL is a research project it was possible to optimize the pilot in order to maximize the understanding of the sequestration process to be gained from the tests. A new injection well was drilled at 150 m from the production well. After completion of the well with casing, cementing and perforations, the perforated zones were tested.

A baseline cross borehole seismic survey was carried out for monitoring purposes in September 2003. Activities in autumn 2003 included the finalizing of the injection facilities. Production has started in the first half of June 2004, to establish a baseline production. First injection tests took place in the first week of July. Once the injection was stabilized, both injection and production were continued until May 2005. During the injection period the process was monitored directly and indirectly to assess any potential, although unlikely, leakage of CO_2 to the surface. The gas that was produced was burned on the site via a flare. The produced water (saline, up to 160 g/l) water was transported to the disposal site of the nearby mine. The fluid level in the production well was regularly monitored, to check the performance of the production pump and to control the pump operation. Along with the field tests, an extensive laboratory program was carried out.

The Upper Silesian Basin (Fig. 4) was selected as the most suitable coal basin in Europe for the application of ECBM. This basin has (relatively) favourable coalbed properties (depth, permeability, gas content, etc.), was subjected to CBM production before. The location of the pilot site in the village Kaniow, about 40 km south of Katowice, was selected at an early stage of the project. There were two wells, 375 m apart, which were formerly used for a short period to produce CBM. The selected site is located within the concession of the "Silesia" (now "Brzeszcze" mine), which has been in operation for ten years. The characteristics of the site have been documented from these activities and from the activities in the nineties by the owner of the existing wells MS-1 and MS-4. As many data as possible were collected and evaluated, in order to get a good background for the development plan of the site. All available documents of former drilling and CBM-production were analysed.

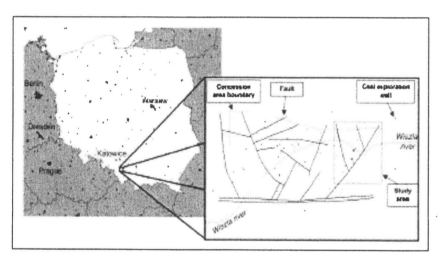

Figure 4. Location of the RECOPOL pilot site

The location of the injection well was determined by the results of the reservoir modelling and the local terrain conditions. A distance of 150 m between the production well and an injection well was supposed to give the best chance of a CO_2 breakthrough within the test period. The injection well was placed down-dip of the production well, and nearly rectangular to the strike of the coal seams. Permits for drilling were arranged and agreement with the landowner was settled. Site construction, mobilization and rig up took place between mid July and August 2003, followed by drilling of the well to a depth of 1,120 m in August and September. Cuttings of the entire interval and core samples of the most important coal seams were taken for various laboratory experiments. Extensive wire-line logging was done before running the 7" casing to determine petrophysical properties, lithology, porosity, saturation, dip, well deviation, etc. The well was completed with casing, cement and perforations.

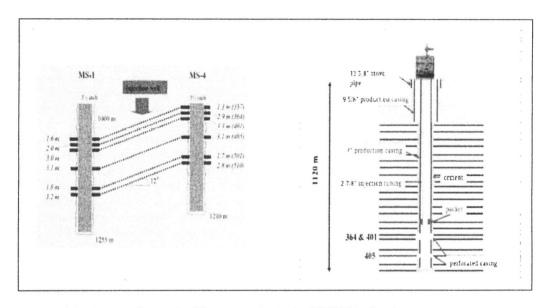

Figure 5 (left). Target coal seams for CO_2 sequestration in the RECOPOL pilot site

Figure 6 (right). Well layout of the injection well MS-3

Figure 7. Overview picture of the RECOPOL injection site, with the wellhead at the front

MONITORING IN RECOPOL PROJECT

The main purpose of monitoring applied during the RECOPOL Project was to control the operations, its effects and safety with special focus on LCO_2 injection and sequestration into the coal seams.

It covered in specific the following:

- LCO_2 injection parameters in the MS-3 well.
- Production parameters in the MS-4 well, with special focus on gas production, its content, mainly considering its volume and concentrations of CH_4, CO_2 and other gases.
- Measurements of gases' content in the soil air, in the shallow 2-meters deep surface wells, as an element of safety and control of gases content changes while conducting the CO_2 injection tests to the coal seams, underlying at the depths below 1000 m (test of breaking through the rock mass).
- Measurements of CH_4/CO_2 contents in the "Silesia" coal mine, considering underground safety in case of increasing content of CO_2 or CH_4 in the ventilated road ways and as an element of control of eventual infiltration of above gases into the isolated galleries – as a result of LCO_2 injection to lower underlying coal seams in the mining acreage of a.m. coal mine.

1. Monitoring of LCO₂ injection parameters in the MS-3 well

Basic element of monitoring in the injection well MS-3 is the SCADA (Supervisory Control And Data Acquisition) system of LCO_2 injection installation equipped with the pressure and temperature sensors (at the CO_2 tanks and the well head), with the flow sensor (on the pipe), working in the digital system (SCADA – ADAM – PC – printer).

2. Production monitoring at MS-4 well

Monitoring was working in the digital-analog system, with the auxiliary manual measurements. It consisted of analog gas and water flow meters, manometers and digital CH_4/CO_2 gas analyser "UL-TRAMAT" connected with the PC system (ADAM – PC – printer). Digital-analog measurements were complemented by manual measurements concerning pH, conductivity (conductometer), temperature and salinity of the water from the MS-4 well, by portable meter type CPC-401 "Elmetron" (PL) as well as chromatographic analysis of gases content and composition in the samples collected periodically with twice a week, and during the period of intensive CO_2 outflow twice a day – one for chromatographic tests and another for the isotopic-ones to define content of durable isotope C^{13} (mass spectrometer).

3. Surface monitoring

Monitoring in the surface 2 meters deep wells was conducted in the digital system, infrared sensors (IR) with the measurement range 0–5% CO_2 were permanently connected with the microprocessor station of monitoring–recording system MSMR-1. Above monitoring was additionally supported by the measurements done by portable infrared gas analyser GA-2000 "Geotechnical Instruments Limited" (U.K.) (O_2 – 21%, CH_4 – 100% i LEL, CO_2 – 100%, atmospheric pressure, temperature) and by chronomatographic gas content and concentration analysis in the soil air samples from the wells.

4. Monitoring in the coal mine workings

It consisted of two methane monitoring & measurement systems type "ZEFIR" and CCT-9U with additional CO_2 sensors type IR manufactured by Carbo-Toka. System "ZEFIR" is taking measurements in the continuous mode, and system CTT-9U with the frequency of 4-minutes. Methane sensors of both systems are located in the workings with the constant ventilation airflow – in the locations pointed out by the ventilation engineer. Additionally as "a backup" 2 samples are being collected from the places located behind the three isolating dams in the area adjacent to the MS-3 well, one is being collected to determine composition and content of gases between the dams, second for determining content of permanent isotope C^{13} (isotope testing at the mass spectrometer).

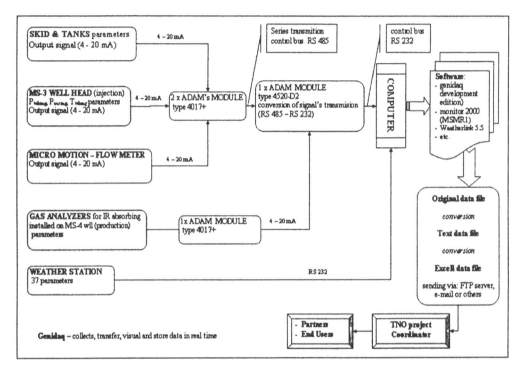

Figure 8. RECOPOL Project monitoring chart

Monitoring in this CO_2 sequestration project is applied for the following reasons: to determine, that the CO_2 is injected into the envisaged coal seams, to ensure that safety and environment are not jeopardised by unforeseen possible leaks, and to improve the coal reservoir behaviour as a consequence of the injected CO_2 and the released coal bed methane. Additionally, monitoring experience is gained in CO_2 sequestration projects with respect to mass balance verification, where the stored CO_2 should equal the injected amount of CO_2 such that the numbers used for emission quota and carbon credits (Kyoto protocol) are correct. Related to the environment and safety issues and the verification of the injection process a general Features, Events and Processes (FEP) analysis is used to identify possible CO_2 leakage to shallower geological formations and the surface. Some threats are listed below.

Examples of FEP's identified [e.g. Hendriks *et al.* 2003; Arts and Winthaegen 2004]:

– Formation damage due to drilling the wells.
– Operational failure of the well.
– Fracturing or fault activation due to increased CO_2 pressure.
– Dissolution or dehydration of seal due to presence of CO_2.
– Unrecognised features in the seal like faults, joints.
– Corrosion of well casing due to CO_2.

The CO_2 is injected in the reservoir in a high-density supercritical phase. The pressure in the production well is low, because the water is pumped out of the well. This creates a pressure gradient from the injection to the production well-bore. The density of the CO_2 will thus lower towards the production well when it is migrating through the coal seams. It is envisaged that no immediate adsorption of the CO_2 with the coal takes place. Therefore, some of the injected CO2 might move upwards until it reaches intra-Carboniferous clay seals and eventually the thick impermeable Miocene shale formation. Local mining experience showed that the Miocene shales are sealing and are not in hydrological contact with the Carboniferous. It is therefore expected, that latter forms a cap rock to the free CO_2, and after CO_2 adsorption also as a seal for the coalbed methane. Based on this analysis monitor techniques are considered, that focus at the wells and at the reservoir and its seal. Additionally, in case there would be leakage through the seal (e.g. along fractures and faults) also the shallower geological formations and the surface need to be monitored.

PREOPERATIONAL AND OPERATIONAL EXPERIMENTS

In order to evaluate possible effects of CO_2 injection, it is mandatory to establish the initial, or baseline, conditions that existed before the start of the injection operations. Cores and cuttings were taken during drilling and analysed in the laboratory to determine the pristine rock conditions. Additionally, an extensive logging suite was measured in the injection well (before it was cased and cemented). After completion of the well, a Cement Bond Log checked the cement job and the water injectivity of the coal seams were tested. The pressure and temperature conditions in the water filled injection well were determined before injection. The initial equilibrium water levels in the casing and tubing were measured. The gas (incl. isotope) and water composition of the production well was analysed to establish the baseline conditions. Baselines were also verified for CO_2 concentrations in the nearby mine and in the soil gas. These baseline conditions will be compared to the measured conditions during and after injection. The baseline measurements are used to determine the reference measurements and possible variations in the measurements. Subsurface models, which include as much available information and data, will be set-up that can represent the observed changes. When this process ends the model is used to predict future behaviour of the coal reservoir and it is determined what monitoring is needed after field abandonment.

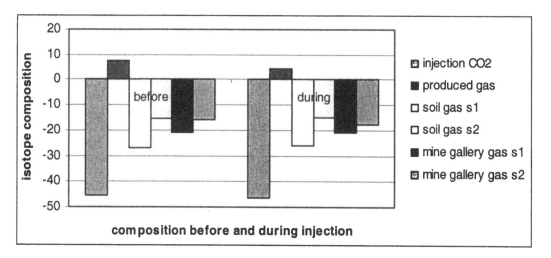

Figure 9. Values of $\delta13$ CO_2 (‰) determined in a period before and after CO_2 injection

The monitoring techniques applied are listed below.

– Isotope analysis: CO_2 has a specific isotopic signature depending on the source gas from which it is generated. Gas samples from the production well, from the near-surface wells and the CO_2 to be injected are taken for isotope signature analyses. Comparing the isotope signatures of the injected and produced gas might give an indication of the breakthrough of the CO_2. In Figure 9 the values of the isotope signatures are displayed. The $\delta 13$ CO_2 signature from the injected CO_2 in the reservoir is distinctly different from the produced CO_2 and from the CO_2 occurring in the soil gas (measured at 2 locations) and in samples from closed galleries from a nearby mine (measured in 2 galleries).

– Soil gas monitoring: IR CO_2 sensors are positioned in 2m deep tubes surrounding the injection well. The local soil thickness, containing biogenic activity, was expected to be less than 2 m. Continuous digital data registration resulted in baseline profiles for the area near the injection well. An example of the registration is shown in Figure 2. Still a variation in concentration is registered, namely (in general) a decrease during the night, with a through in about the middle of the day, and an increase during the rest of the day. This variation might be caused by soil biogenic activity of e.g. CO_2 emission by roots and microorganisms through respiration and on the other hand decrease due to photosynthesis by plants. Also the weather (e.g. wind, rain water) and nearby industrial activities are expected to influence the measured concentration. Seasonal effects are also observed.

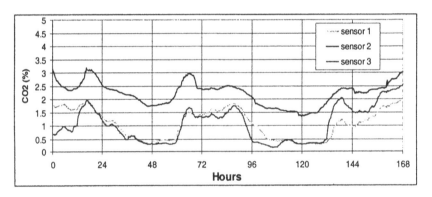

Figure 10. Baseline soil gas measurements, over a 7-day period in April before CO_2 injection, indicating the variation in CO_2 concentration

– Mine gallery gas monitoring: Baseline CO_2 concentration measurements are taken from galleries in a nearby Silesia coal mine, at a depth of 270 m and at a horizontal distance of several hundreds meters of the site. These CO_2 measurements acquired in these mine galleries show a large variation in concentrations up to a maximum of circa 1%. In sealed galleries CO_2 concentrations are measured up to 10%, mostly as a result of oxidation of coal and natural content in coal.

– Composition of the produced water: In order to detect possible compositional changes in the production water, baseline measurements are performed to determine the composition and its natural variation. This analysis showed that the water is highly saline. It is expected that the concentration of Ca^{2+} and Mg^{2+} will increase as a result of CO_2 injection, because the formation water is expected to acidify when coming into contact with CO_2.

– The pressure and temperature of the tubing and the pressure in the casing of the injection well are continuously monitored. Leaking of the tubing would be detected in an early stage by in increase in casing pressure.

CONCLUSIONS

Monitoring of CO_2 injection was performed:
- To improve the understanding of CO_2 storage in these coal layers.
- To verify, that safety and environment are not jeopardised.
- To determine, that the CO_2 is injected into the intended coal layers.

This monitoring programme is based on FEP analyses and aims at different parts of the subsurface, the surface and the wells. Monitoring results, e.g. from soil gas, composition of produced gas and water, cross-well seismic acquisition and isotope signature analysis, are discussed. These measurements should significantly contribute to the development of a subsurface model, that predicts future behaviour of the stored CO_2 and the coal after field abandonment.

KEYWORDS

CBM – Coal Bed Methane;

EOR – Enhanced Oil Recovery;

ECBM – Enhanced Coal Bed Methane;

FEP – Features, Events and Processes Analysis;

LCO2 – Liquid CO_2;

LEL – Lower Explosion Limit;

GA-2000 – Infra-red Gas Analyser – (O_2 up to 25%, CH_4 up to 100% and LEL, CO_2 up to 100%, atmospheric pressure and temperature);

GA-94 – Infra-red Gas Analyser. Measures concentrations of landfill gases CH_4, CO_2 and O_2 (reading range: CH_4: 0 – 100%, CO_2: 0 – 50%, O_2: 0 – 21%; pressure range: 900 to 1100 mbar). Automatic recording of time and date with each stored data. Data can be downloaded to PC;

ADAM-4017 is a 16-bit, 8-channel analog input module that provides programmable input ranges on all channels. This module is an extremely cost-effective solution for industrial measurement and monitoring applications. Its opto-isolated inputs provide 3000 VDC of isolation between the analog input and the module, protecting the module and peripherals from damage due to high input-line voltages. ADAM-4017 offers signal conditioning, A/D conversion, ranging and RS-485 digital communication functions. The module protects equipment from ground loops and power surges by providing opto-isolation of A/D input and transformer based isolation up to 3000 VDC. The ADAM-4017 uses a 16-bit microprocessor-controlled sigma-delta A/D converter to convert sensor voltage or current into digital data. The digital data is then translated into engineering units. When prompted by the host computer, the module sends the data to the host through a standard RS-485 interface;

MSMR-1 – station of the microprocessor monitoring-recording system. It is stationary system for measuring: explosive gases concentration, oxygen and toxic gases concentration and other no electrical values. This system is adapter for external instruments control, using programmable transmitters. It has internal data memory for results recording;

ZEFIR and CTT-63/40U – these are two methane measurements systems equipped with appropriate software, which are used by "Silesia" coalmine to provide constant monitoring of gas parameters in the mine workings (CH_4, CO, O_2, etc.), together with additional CO_2 sensors type IR manufactured by CarboToka, with the measurement range 0–5%. "ZEFIR" and "CTT-63/40U" systems are working continuously, and CO_2 sensors on a cyclic basis with the 4 minutes frequency;

ULTRAMAT 23 – Gas analysers for IR-absorbing gases (ULTRAMAT 23) measure 2 gas components at once: CO_2, CH_4. We are using ULTRAMAT 23 to measure CO_2 in range 0–5 or 0–25% and to measure CH_4 in range 0–20% or 0–100%.

Acknowledgements

The RECOPOL was carried by an international consortium of research institutes, universities and industrial partners. The project is funded by the EC (ENK-CT-2001-00539). Shell International, JCoal, the Federal Region of Wallonie (through the Faculté Polytechnique de Mons) and the Polish and Dutch governments (via Novem) are thanked for their support.

REFERENCES

Van Bergen F., Winthaegen P., Pagnier H., Jura B., Kobiela Z., Skiba J.: Monitoring Techniques Applied for CO_2 Injection in Coal. 67th European Association of Geoscientists & Engineers Conference, Madrid, Spain, 13–16 June 2005 (Conference Proceedings).
Cybulski K., Jura B., Krzystolik P., Skiba J.: CO_2 Sequestration in the Geological Formations – a Chance for the Future of Clean Coal Technologies. Conference Concerning Application of Clean Coal Technologies, Katowice, GIG, 18 May 2005.
De-Smedt G., Hasanov V., Jura B., Kretzschmar H.J., Krzystolik P., Müller-Syring G., Pagnier H., Skiba J., van Bergen F., Wentink P.: Recopol – Site Development: MS-3 Injection. Scientific Workshop "Recopol", Szczyrk 10/11.03.2005, Internet edition: http://recopol.nitg.tno.nl
Adamaszek Z., Jura B., Krzystolik P., Pagnier H., Skiba J., van Bergen F.: Recopol – Site Development: MS-4 Production. Scientific Workshop "Recopol", Szczyrk 10/11.03.2005, Internet edition: http://recopol.nitg.tno.nl
Krzystolik P., Cybulski K., Skiba J.: Podziemne magazynowanie dwutlenku węgla – strategiczną szansą rozwoju górnictwa węgla kamiennego. Szkoła Eksploatacji Podziemnej, Szczyrk 21.02.2005.
Bruining H., Bossie-Codreanu D., Busch A., Choi X., De-Smedt G., Fröbel J., Gale J., Grabowski D., Hadro J., Hurtevent D., Jura B., Kretzschmar H.J., Krooss B., Krzystolik P., Mazumder S., Müller-Syring G., Pagnier H., Reeves S., Skiba J., Stevens S., van Bergen F., van der Meer B., Wentink P., Winthaegen P., Wolf K.H.: Recopol – Field Experiment of ECBM-Co_2 in the Upper Silesian Basin of Poland. 7th International Conference on Greenhouse Gas Control Technologies, Vancouver, Canada, 05–09 September 2004.
Arts R. and Winthaegen P. 2004: Monitoring Options for CO_2 Storage. Carbon Capture Project, Technical Publication (in press).
Hendriks C., Wildenborg T., Feron P., Graus W. and Brandsma R. 2003: EC-Case Carbon Dioxide Sequestration. Report M 70066, December 2003.
Krzystolik P., Mizera A., Skiba J.: Polskie metody pozyskiwania metanu z kopalń węgla kamiennego i jego utylizacja. International Coal Bed Methane Symposium, Tuscaloosa, Alabama 05.2003. Symposium Proceedings.
Van Bergen F., Pagnier H.J.M., van der Meer L.G.H., van den Belt F.J.G., Winthaegen P.L.A. and Westerhoff R.S. 2002: Development of a Field Experiment of CO_2 Storage in Coals Seams in the Upper Silesian Basin of Poland (RECOPOL). Proc. of GHGT-6, Kyoto, Japan.

International Mining Forum 2006, Sobczyk & Kicki (eds) © 2006 Taylor & Francis Group, London, ISBN 0415 401178

Fly Ash Suspension with CO_2 as a New Method of Gob Fire Prevention in Coalmines

Wacław Dziurzyński
Strata Mechanics Research Institute of the Polish Academy of Sciences. Krakow, Poland

Radosław Pomykała
AGH – University of Science and Technology, Faculty of Mining and Geoengineering.
Krakow, Poland. e-mail: rpomyk@agh.edu.pl

ABSTRACT: The new method of spontaneous gob fire prevention in underground coalmine was described in the paper. It is the fly ash suspension together with carbon dioxide application. The basic aim of the application was to limit fire hazard (spontaneous heating processes elimination) by decreasing the airflow through the gob (suspensions) and inertization of the gob atmosphere (carbon dioxide).

In order to evaluate the efficiency of using the new prevention method, in-situ tests have been conducted during the normal coal exploitation of a longwall system. The second part of the tests concerned the evaluation of the gob inertization influence on the carbon dioxide concentration in the outlet stream of air. The results of the tests were presented.

KEYWORDS: Underground coal mining, spontaneous heating and fire, fire prevention, inertization, fly ash suspension

1. INTRODUCTION – A FEW ASPECTS OF GOB FIRE PREVENTION

Exploitation conditions of hard coal in Polish underground mines are hard. Fire hazard is one of the most important problems. It is connected with a natural feature — propensity of coal to oxidation in a low temperature, which results in spontaneous heating and in consequence — fire. Three factors are necessary to start a spontaneous fire: combustible material (for example: crushed coal), the presence of oxygen and conditions for heat accumulation. The place where the spontaneous fire hazard is especially high is a longwall gob.

New methods and means for fire hazard fighting have been researched and developed for many years. Thanks to the high level of ventilation services in mines and the efficient early detection of fire hazard the special emphasis can be put on gob fire prevention. The limitation of airflow through the gob is actually the most popular direction of gob fire prevention. One of the possible ways is a good isolation another one is elimination of oxygen from the gob atmosphere by inertization (replacement of oxygen by inert gas).

The suspension technology is widely used for the gob isolation. This technology consists of gravitational hydraulic transport of fine-grained energy wastes (a very popular term: fly ash suspension) or energy and mine wastes into the gob. The fly ash suspensions have been used in the Polish underground coalmines for twenty years. They cause caulking, isolation and cooling of a longwall gob.

Until recently such gases as nitrogen, carbon dioxide or cooled combustion gases, have been used to fight spontaneous fire in a developed phase, they worked as an assistance of ventilation methods. Nowadays inert gases are more often used as prevention means.

The application of the fly ash suspensions with a carbon dioxide is an idea, which appeared while searching for a new method of fire prevention, especially the one of an effective application of inert gas to spontaneous heating places. The basic aim of the application was to limit fire hazard (spontaneous heating processes elimination) by decreasing the airflow through the gob (suspensions) and inertization of the gob atmosphere (carbon dioxide). The synergy effect may result from the method.

The main purposes of suspensions – caulking, isolation and cooling of a longwall gob – are still realised. Additionally it is possible to applicate inert gas – carbon dioxide – into the gob, without an excessive installation development. The flow of carbon dioxide in the gob is supported by the suitable longwall geometry.

2. THE RANGE OF THE TESTS

In order to evaluate the efficiency of using a fly ash suspension with carbon dioxide as a new prevention method, in-situ tests have been conducted during the normal coal exploitation of a longwall system. Two spontaneous fire indicators were suggested for the evaluation: the carbon monoxide concentration (CO [%]) and the Graham indicator (G).

The last one is defined as a ratio of the increase of carbon monoxide concentration to the decrease of oxygen concentration (O_2 [%]), according to the formula:

$$G = \frac{CO}{0,265 N_2 - O_2}.$$

where N_2 – nitrogen concentration [%].

For the calculation of the indicator level, the regular measurements of the air composition in different points of the longwall were taken while the fly ash suspensions with carbon dioxide were being supplied to the gob.

The second part of the tests concerned the evaluation of the gob inertization influence on the carbon dioxide concentration in the outlet stream of air. It is especially important in view of mining regulations, which allow for a 1% concentration of CO_2.

3. TECHNOLOGY OF FLY ASH SUSPENSION WITH CARBON DIOXIDE PRODUCTION

The suspensions are produced in a special installation (scheme on Figure 1). The energy wastes are delivered from the "Rybnik" Power Plant to the mine by the rail transport and pneumatically unloaded into the cisterns. The preparation of the suspensions is realised in the closed mixer, where the ashes and water are added. While the ready suspensions flow to the shaft's backfill pipeline, the gaseous carbon dioxide is added. Now the hydraulic transport of mixtures into the gob begins.

The suspensions were delivered from the backfill pipeline, end parts of which were left into the gob – in the liquidated upper gateway. The distance between the end of the pipeline and the longwall head varied from a few to dozens meters, it was caused by local conditions. After some time the end of the pipeline got plugged and the end parts of the pipeline were disconnected in an accessible place in the upper gateway.

Carbon dioxide was delivered to the mine in a liquid phase and stored in the tank of 30 tonnes capacity. The gasification of CO_2 was conducted by an evaporating dish and an electric heating unit. The efficiency of gasification installation reached 4,5 m^3/min.

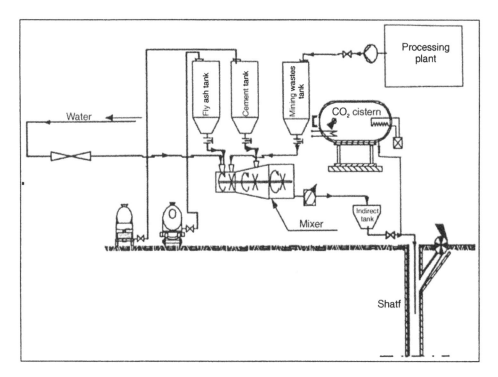

Figure 1. Scheme of the fly ash suspension with carbon dioxide production installation.
(Source: "Borynia" Coalmine)

4. THE PLACE OF THE TESTS

The in-situ tests have been conducted in the longwall D-31 in the "Borynia" Coalmine, which be-longs to the Jastrzebie Coal Company in Southern Poland. The tests started in May 2004, and fi-nished in April 2005.

The fly ash suspensions with carbon dioxide were supplied to the gob of the D-31 longwall, which led in the upper layer of the 405/1 stratum. The length of the longwall [1] was from 140 m in the beginning to 210 m in the end, the planned advance – 600 m and the average height – 3 m. The lowest point of the longwall was located at the beginning of the lower gateway – D-31 (first part of the gob), the highest points – at the end of the upper gateway – D-31a. Both the gateways were li-quidated behind the head of the longwall.

The longwall geometry helped to the carbon dioxide in migration into deeper parts of the gob.

The analyses of the air composition (especially concentration of oxygen, carbon dioxide, car-bon monoxide, nitrogen and methane) were carried out in measurement stations, which location is showed in Figure 2. On this basis the Graham indicator and the carbon monoxide indicator were calculated.

Denotation and location of measurement station:
- IN – inlet airflow;
- S10, S20 and S70 – station located behind 10, 20 and 70 section of the hydraulic support;
- FS – first section of the hydraulic support (crossing of the head of the longwall and the upper gateway);
- GL – D-31 upper gateway liquidation;
- OU – outlet of the airflow.

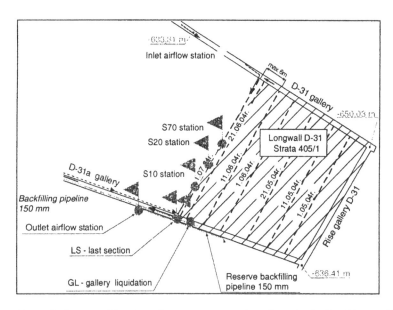

Figure 2. Scheme of location of measurement station (flags) in the D-31 longwall.
(Source: "Borynia" Coalmine)

5. RESULTS – THE EVALUATION OF THE FIRE PREVENTION METHOD

The frequency of application of the fly ash suspensions with carbon dioxide to the gob was very irregular. It depended on spontaneous heating development, and was in accordance with the value of fire indicators. During the tests there were some periods, when the heating process in the gob advanced. Then the fire prevention intensified – the quantity and frequency of the fly ash suspension with carbon dioxide supplied to the gob increased.

In Figure 3 the data concerning the quantity of supplied CO_2 is shown, and in Figure 4 – the quantity of supplied fly ash as a part of suspensions.

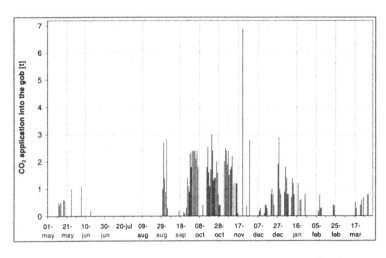

Figure 3. Application of the carbon dioxide into the D-31 longwall gob

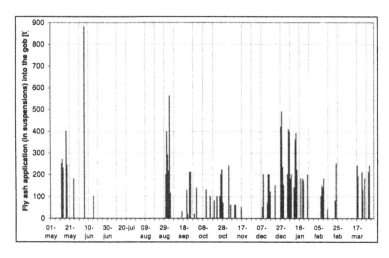

Figure 4. Application of the fly ash, as a part of suspensions, into the D-31 longwall gob

There was a period between September and November 2005, when heating process in the gob advanced, and the situation was serious. The evaluation of the use of the fly ash suspensions with carbon dioxide as a fire prevention method focused on this period. The values of fire indicators (CO and G) and the amount of carbon dioxide (transported with suspension) applicated into the gob during this period, measured in the FS station, are presented in Figures 5 and 6. In the beginning of September the value of the Graham indicator exceeded 40 $[10^{-4}]$. After the supply of 9 tonnes of CO_2 and about 1800 tonnes of fly ashes (into the suspensions) within six days the Graham indicator value was decreased to 20 $[10^{-4}]$. The situation was stable, but for a few days only. Between 21st September and 6th October 24 tons of the carbon dioxide and 730 tons of fly ash were supplied to the D-31 longwall gob. Thanks to that the decrease of the value of the fire indicators could have been observed again. And again the situation was stable for a few days. This scenario repeated twice, and the value of the Graham indicator reached more than 120 $[10^{-4}]$, and after a few days of application of the fly ash suspension with CO_2 (22÷24 Mg CO_2 was supplied) was decreasing to 20÷30 $[10^{-4}]$ level. The situation was under full control before the end of November 2004.

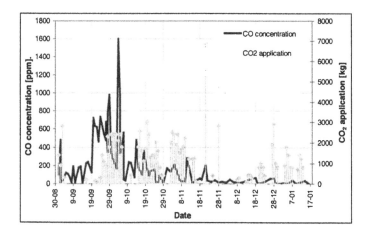

Figure 5. CO concentration measured in FS station and the amount of CO_2 supplied to the D-31 longwall gob, in the described period

111

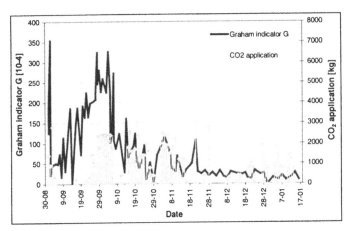

Figure 6. The value of the Graham indicator (computed on the base of measurements in FS station) and the amount of CO_2 supplied to the D-31 longwall gob, in the described period

At this time the second part of the tests began. It was a single application of carbon dioxide with maximum efficiency of the installation. It caused a decrease of value of Graham indicator. This action and later regular of the fly ash application with carbon dioxide supplying helped to stabilise the situation (the level of the G value no more than 40 [10^{-4}]) until April 2005. When the value of the Graham indicator reached 30÷40 [10^{-4}] it was a signal to start suspensions with CO_2 start application.

6. INFLUENCE OF CARBON DIOXIDE APPLICATION UPON THE MINE ATMOSPHERE

During the test in the D-31 longwall, the permanent control of the mine air composition, in front of and behind the head, was conducted. In few cases an increasing value of CO_2 was noted, especially in FS and GL stations, when heating processes in the gob went on. The concentration of carbon dioxide in the outlet airflow never reached the permissible level (1%).

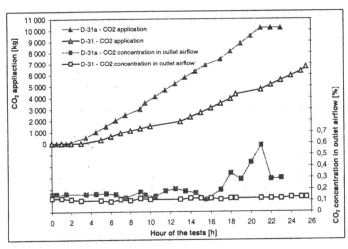

Figure 7. The quantities of carbon dioxide applicated into the gob during the tests (upper graphs, left scale) and CO2 concentration measured in the outlet airflow – OU station (lower graphs, right scale)

Special tests were carried out in order to precisely evaluate the CO_2 application to the gob and on the carbon dioxide concentration in the outlet airflow. The first of them, mentioned above, was realized on 21st November 2004. About 6.9 tonnes of CO_2 was added into the gob within 21 hours by the pipeline in the liquidated part of the lower gateway D-31. It is significant that the direction of CO_2 applications was in accordance with the airflow direction through the gob (from the lower gateway D-31 to the upper gateway D-31a). The concentration of the CO_2 was measured in the OU station in the upper gateway D-31a. The results of the tests are presented in Figure 7. The increase of the CO_2 concentration was not noted.

The second test was conducted in April 2005, when 10.2 tonnes of CO_2 was added to the gob within 20 hours from the pipeline in the upper gateway D-31a. The pipeline in the liquidated part of the upper gateway was 55 m long. In that case the direction of the CO_2 application was opposite to the direction of the airflow through the gob. The measurements taken in the OU station showed an increase in the values of CO_2 concentration in comparison with the results of the previous test. The permissible level of CO_2 – 1% – concentration was not reached.

7. SUMMARY

The example of a new method of the fire prevention, the application of fly ash suspension with carbon dioxide during regular longwall exploitation in an underground coalmine was described. Several times higher values of fire indicators (CO concentration and the Graham indicator) were observed. It pointed at spontaneous heating in the gob. The new method allowed for retarding and controlling all the processes without disturbing the exploitation.

During the research the limitation of heating process was observed (abating of heating symptoms) especially after the periods of intensified application of fly ash suspension with carbon dioxide. The best results were obtained when stable amounts of fly ash suspension with carbon dioxide were regularly applied. Longer breaks, even several days, caused reappearance of spontaneous heating process.

A good example of this is the action undertaken from September 2004 to November 2004. After the application of fly ash suspension with carbon dioxide within this period, there was a long period without spontaneous heating symptoms.

The conducted research indicates that it is possible to avoid exceedance of permissible level of CO_2 concentration in the airflow even if a maximum dose of CO_2 was applied. It may be stated that, according to assumptions, CO_2 migrates into deeper parts of the gob. At the same time the influence of the place of CO_2 application on its flow in the gob was observed.

REFERENCES

Budryk W.: Górnictwo tom XI: Pożary i wybuchy w kopalniach. Wyd. Górniczo-Hutnicze, Stalinogród 1956.
Mazurkiewicz M., Piotrowski Z., Tajduś A.: Lokowanie odpadów w kopalniach podziemnych. Biblioteka SEP, Kraków 1997.
Mazurkiewicz M., Piotrowski Z., Pomykała R.: Zawiesina popiołowo-wodna jako środek transportu CO_2 do zrobów kopalni podziemnej. Materiały Konferencyjne SGO, Kraków 2004.
Mazurkiewicz M., Piotrowski Z., Pomykała R.: Zawiesina popiołowo-wodna z dodatkiem CO_2 jako środek w profilaktyce przeciwpożarowej. Materiały Konferencyjne SEP, Szczyrk 2005.
Pomykała R.: Właściwości zawiesin popiołowo-wodnych z ditlenkiem węgla stosowanych w profilaktyce pożarowej, Praca doktorska, niepublikowana, Kraków 2006.
Strumiński A.: Zwalczanie pożarów w kopalniach głębinowych. Wyd. Śląsk, Katowice 1996.
Tor A. et al.: Doświadczenia w inertyzacji z zastosowaniem dwutlenku węgla do zwalczania zagrożenia pożarowego w kopalniach Jastrzębskiej Spółki Węglowej S.A. Materiały 3 SAG, Zakopane 2004.

International Mining Forum 2006, Sobczyk & Kicki (eds) © 2006 Taylor & Francis Group, London, ISBN 0415 401178

A New Method for Assessing the Status of Sealed-off Coal Mine Fires

Yuan Shujie
Anhui University of Science and Technology. Huainan, Anhui, China

Nikodem Szlązak, Dariusz Obracaj
AGH – University of Science and Technology. Cracow, Poland

Si Chunfeng
Wanbei Coal – Electricity Company. Suzhou, Anhui, China

ABSTRACT: The paper presents a method to assess the status of sealed-off underground coal mine fires. The assessment is based on the analysis of gas samples taken from the fire areas. Considering the mining an geological factors influencing the results, the authors suggested their own method of analysing and assessing the condition of a sealed-off fire with the use of so-called "fire characteristic", which graphically describes the variations of five fire gas components – oxygen, nitrogen, carbon dioxide, carbon monoxide and hydrocarbons – in time. An analysis of the fire gas components' tendencies and their correlations allows drawing proper conclusions about the fire status. In order to mark fire indicators' tendencies in time the authors applied the time series analysis. The cases studied confirmed that the suggested method is beneficial for analysing and assessing the states of fires located in sealed-off areas.

KEYWORDS: Coal mines, mine fires, spontaneous combustion, sealed-off fire areas, fire status assessment, fire indicators

1. INTRODUCTION

Underground fires are one of the serious hazards in coal mines. Fire fighting is often done by sealing off the area containing the fire. After the fire area is sealed, the necessity arises to assess the fire status, in order to decide whether to open the sealed area. Heterogeneity of the medium makes it difficult to determinate the physical phenomena occurring in the sealed space. The methods of fire status assessment depend mainly on gas samples taken from the sealed area and analyses of their composition. The various methods of fire status assessment are based on fire indicators, which are calculated from the concentrations of fire gases in the samples. Known fire indicators are not always correctly used and not always describe the status of the fire contained in a sealed area properly. The changes of fire indicators' values in time have more significance in the proper interpretation of fire state. Up to now no reliable, practical indicators signalising extinction of a sealed-off fire has been worked out. A fire may exist in a sealed area from several weeks to several decades. Therefore, proper interpretation of fire status parameters is very important in the safe opening of sealed areas as soon as possible and recovering the locked up coal reserves. Although a range scientific investigations were done, there are no unambiguous criteria for assessing the fire status and the problem is still not solved.

2. FACTORS INFLUENCING SEALED-OFF FIRE STATUS ASSESSMENT

Fires in coal mines can be divided into endogenous and exogenous, according to their genesis. Endogenous fires are caused by spontaneous combustion, and exogenous fires by open flame. In the case of tight sealing of fire areas both of the two kinds of fires go through the same process leading to their extinction. In the process of fire extinction, if oxygen is present in a sealed area and temperature higher than normal atmospheric temperature, coal spontaneous heating or combustion may occur.

The necessary condition for spontaneous coal combustion to occur is concurrent presence of the following [Strumiński 1996, Maciejasz 1977]:
– Fragmentised coal prone to low-temperature oxidization.
– Appropriate airflow.
– Heat accumulation during coal oxidization.

The first of the conditions is connected with the natural character of coal, and the other two depend on mining-technological conditions, namely the mining system, ventilation method and so on. Except for its natural tendency to spontaneous combustion, the following physical characteristics of coal also affect its spontaneous combustion [Cygankiewicz 2000, Strumiński 1996]:
– Degree of metamorphism.
– Petrography composition.
– Moisture content.
– Initial temperature.
– Fragmentation.
– Thickness of coalbed.
– Coalbed's dip angle.

For certain coal types the main factor influencing spontaneous combustion is physical parameters of the air flowing through the fragmentised coal pile, for example flow speed, temperature and humidity, which are affected by mining-geological conditions: exploitation system, ventilation system and strength parameters of rock. A substantial number of theories were formulated on spontaneous combustion of coal [Maciejasz 1977, Strumiński 1996, Strzemiński 1984].

Determination of sealed fire extinction criteria is very important for safe opening of sealed areas and resuming ventilation there. At the moment of opening of sealed areas spontaneous heating or combustion often re-starts, and sometimes explosion of fire gas occurs. Fire extinction depends on oxygen concentration in the atmosphere and coal temperature. The two parameters don't change identically with time. Minimum concentration of oxygen supporting flaming combustion is shown in Tables 2.1 and 2.2.

Table 2.1. Minimum concentration of oxygen supporting flaming combustion of some combustible gases [Greuer 1973]

Combustible gas	Other non-combustible gases in mixture	
	Air and nitrogen	Air and carbon dioxide
CH_4	12,1	14,6
H_2	4,9	5,9
CO	5,6	5,9

Table 2.2. Minimum concentration of oxygen supporting flaming combustion of some fuels [Mitcher 1996]

Fuels	Timber	Low rank coal	Middle rank coal	High rank coal	Methane
Ignition	2	5	10	18	12
Flaming combustion	2	5	5	5	12
Smouldering	0	0	0	0	-

The minimum concentration of oxygen supporting combustion is far smaller for solid fuels than for combustible gases. Research shows that the minimum concentration of oxygen supporting combustion of coal and timber is 2% [Mason 1993, Stark 1965]. In general, it's accepted that combustion of solid fuels stops when concentration of oxygen falls below 2%.

The main reason of wrong decisions to open sealed fire areas is incorrect analysis and assessment of fire status.

Wrong assessment of fire status may be caused by following factors:

– The seal doesn't completely stop air from flowing through the fire area, the air carries oxygen to the fire source and allows spontaneous heating to progress.
– Because of carbon monoxide adsorption by coke it can't be unambiguously said that fire is extinguished only because the concentration of carbon monoxide decreased to zero.
– Pressure equalization on stoppings disables airflow through them. In sealed areas small amount of carbon monoxide may remain for a long time. It may be produced by low-temperature oxidation of other organic substances, not only coal combustion.
– In case of untight stoppings the concentration of oxygen may increase with the concentrations of CO, CH_4, H_2, C_xH_y and CO_2 decreasing. If the sealed area is situated near a big fault and close to the surface, the concentration of oxygen in the atmosphere in the sealed area may stay high.
– Methane inflow from strata or nitrogen supply to the fire source may change the composition of fire gases and influence proper assessment of fire status on the basis of gas samples taken from the sealed area.
– Because of unknown temperature distribution in a sealed area it's difficult to estimate the fire source temperature. If the temperature remains high, smouldering may occur even at very small concentration of oxygen, which may cause fire to re-start when the sealed area is opened.

The above shows that we can't properly assess fire status on the basis of a single indicator neither by its spot value, nor its changes in time. To avoid mining and geological factors influencing the results, fire status assessment can be done by analysing the fire gas composition, the correlation between its component elements and the time trends.

3. METHOD OF PROCESSING DATA FROM SEALED-OFF FIRE AREAS

Spontaneous heating or combustion in sealed-off areas is a slow, sequential process of coal oxidation. Fire extinction is also a sequential process. Sudden changes of gas composition can't be regarded as a basis for assessment of a sealed-off fire status. We can analyse fire state by studying changes of gas concentration in some period of time by taking a series of samples. In order to avoid drawing incorrect conclusions regarding the changes of fire gas composition caused by mining-geological reasons or inaccurate analyses of gas samples, "three period moving average" method is used to analyse the tendency of fire gas composition in time.

As a result of serial (at certain time intervals) surveying of concentration of gases behind stoppings we can get some number of measurement values. When taking a large number of measurements certain statistical dispersion of results always appears. The results can be regarded as random variables. The changes of gas concentration may result in difficulties in interpreting the phenomena occurring in the sealed area. As an example broken line in Figure 3.1 shows oxygen concentration changes in a sealed area. On its basis it's difficult to estimate tendency of gas concentration, decreasing or increasing, in time. In this case, time series analysis can be applied to describe the course of gas concentration changes.

Time series analysis can be applied in the case of unknown functional relationship between random variables y and time intervals t [Brandt 1999]. Time series can be defined as series of random variables with connected distribution $\{y_i; t = 1... n\}$. Results of gas composition measurement and fire indicators calculated on their basis are the variables. Random variables are the sum of true measurement values η_i and measurement errors ε_i. If random variable function is subject to certain trend, it can be described by a curve passing the measurement points. In order to obtain possibly smooth function of variable t, every value can be written as:

$$u_i = \frac{1}{2k+1} \sum_{j=i-k}^{i+k} y_i \qquad (3.1)$$

It's an average (unweighted) value of measurements taken in moments t_{i-k}, t_{i-k+1}... t_i, t_{i+1}... t_{i+k}, and it's named as moving average of variable y_i. Variable k presents the time interval, on which values of variable y_i are averaged. When k = 1 the formula (3.1) yields three period averages, when k = 2 – five period averages, and etc. For the examined case of oxygen concentration shown in Figure 3.1, their three period averages are calculated and presented in Figure 3.1 by solid line. On the basis of changes of gas components' concentrations and fire indicators' values by three period average, we can easily interpret their trends in time. In the following analysis the three period average method will be used.

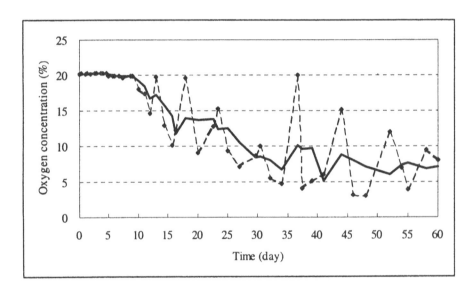

Figure 3.1. Oxygen concentration changes in a sealed area.
--- ♦ --- trend on basis of measured value of oxygen concentration,
— trend on basis of three period average value of oxygen concentration

4. APPLICATION OF FIRE CHARACTERISTIC TO ANALYSE SEALED-OFF FIRE STATE

4.1. *Idea of fire characteristic*

A general analysis of fire tendency (development or extinction) can be supported by "fire characteristic", which in the same figure graphically describes the concentration tendencies of the five components of fire gas – oxygen, nitrogen, carbon dioxide, carbon monoxide and hydrocarbons – in time. Development trends of particular components of fire gas in an examined period can show useful information.

Change tendency of oxygen concentration in sealed areas. If coal oxidation in sealed areas occurs, oxygen concentration decreases. The rate of oxygen concentration decrease depends on oxidation intensity. The bigger the oxidation intensity, the greater the rate of oxygen concentration decrease. In case of spontaneous heating or combustion of coal, oxygen concentration fluently decreases, and its course can be described by exponential function in dependence on the time from the moment of sealing the fire area. When explosion of gas or coal dust occurred, oxygen concentration in the sealed area intensively decreases. If spontaneous heating, combustion or explosion doesn't occur, course of oxygen concentration changes is monotonous. If fire is subject to extinction, oxygen concentration in the atmosphere increases or stabilizes around a certain value. When gases, such as methane from strata, inflow to the sealed area, oxygen concentration decreases.

Change tendency of carbon monoxide concentration. Carbon monoxide is the main fire indicator gas. The rate of carbon monoxide concentration increase in a sealed area is considerably greater than that of other gaseous products of coal oxidation. Carbon monoxide concentration increase indicates combustion or low-temperature oxidation of coal. As opposed to carbon dioxide, the origin of carbon monoxide is only combustion or low-temperature oxidation of coal and other organic substances. The lower limit of carbon monoxide concentration in the atmosphere below which it does not burn in fire is 12.5%. It is almost not changed by other reasons except fire. Generally after fire extinction, carbon monoxide concentration in atmosphere slowly gradually decreases. In some cases, carbon monoxide concentration in sealed areas may remain high for a long time after its extinction, which may be the reason for erroneous assessment of the fire status.

Change tendency of carbon dioxide concentration. Carbon dioxide is the main gas produced during coal combustion. Carbon dioxide production is higher than that of other gaseous combustion products. During a fire carbon dioxide concentration increases with oxygen concentration decrease at the same or higher rate. Inversely, if carbon dioxide concentration in the sealed area atmosphere decreases, oxygen concentration will increase. If oxygen concentration decreases slowly, the intensity of fire development is not big. The fact of small rate of carbon dioxide concentration decrease can't be the basis to state that the intensity of fire development is small. In coal mines carbon dioxide may come from strata or be produced in a reaction of carbonates with acid water. Carbon dioxide can be adsorbed by coke or soot produced in the fire and be dissolved in water.

Change tendency of nitrogen concentration. If the process of spontaneous heating in a sealed area doesn't occur, the tendency of nitrogen concentration change is similar to that of oxygen. When the process of spontaneous heating of coal occurs, the tendency of nitrogen concentration changes is contrary to that of oxygen-nitrogen concentration increases and oxygen concentration decreases. When open fire exists, nitrogen concentration remains stable or slowly decreases, while oxygen concentration decreases fast. When explosion occurred, nitrogen concentration remains around its normal concentration in the air, and oxygen concentration intensively decreases.

Change tendencies of hydrogen and hydrocarbons concentrations. Hydrogen (H_2), ethane (C_2H_4), propylene (C_3H_6) and sometimes acetylene (C_2H_2) or butylene (C_4H_8) release in certain sequence with the increase of coal temperature. Detection of hydrogen, ethylene, propylene and acetylene in gas samples in certain sequence is the sign of spontaneous heating development, while during reduction in spontaneous heating the gases disappear in reverse sequence.

Change tendencies of fire indicators in time supply better information about the processes taking place in sealed areas than their spot values.

On the basis of results from general analysis of fire development according to fire characteristic, we can farther more analyse the course of complex fire indicators changes.

4.2. An example of assessing a sealed-off fire status with the use of fire characteristic

In "X-Q" coal mine longwall 1123(3) area was sealed off from the ventilation system by upper and lower stoppings because of spontaneous heating of coal. After sealing of the fire area gaseous nitrogen was injected into the sealed area from the side of the lower stopping to create inert atmosphere. Gas samples were taken from behind the stoppings. On the basis of gas composition analyses fire characteristic were obtained (Figures 4.1 and 4.2).

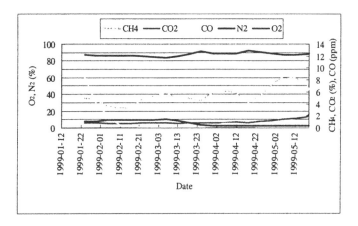

Figure 4.1. Fire characteristic obtained on the basis of gas samples taken from behind the upper stopping

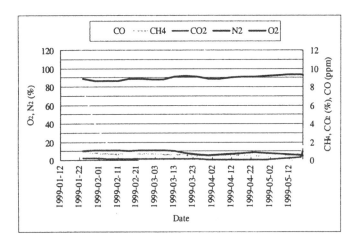

Figure 4.2. Fire characteristic obtained on the basis of gas samples taken from behind the lower stopping

General analysis of fire characteristics on the basis of gas samples from behind both of the stoppings shows that the change tendencies of carbon monoxide concentrations behind the two stoppings are identical. After sealing off the longwall 1121(3) carbon monoxide concentration decreased fast and on January 21, 1999 reached zero. The maximum value of carbon monoxide concentration was only 13 ppm, which probably is the result of dilution by nitrogen injected into the sealed area. In this situation we can't estimate the status of spontaneous heating or combustion of coal only on basis of carbon monoxide concentration value. A comparison of carbon monoxide concentration trend with the trends exhibited by the other components of fire gas can be helpful in fire status estimation. In Figure 4.2 can be seen that nitrogen concentration is about 90% and at the end of the period it presents increasing trend. Meanwhile oxygen concentration is about 10%, and then decreases to about 7%. Carbon dioxide concentration is below 1%, and methane concentration below 2%. Methane and carbon dioxide concentrations have decreasing trends similar to that of oxygen, which indicates that their decrease is a result of dilution by nitrogen injected into the sealed area, and oxygen decrease is not due to oxidation of coal. After January 21, carbon monoxide disappeared, which can indicate, that from this day oxidation process stopped. Before January 21, 1999 even though carbon monoxide appeared oxygen concentration remained at a level of about 10% and didn't show decreasing trend. The course of oxygen, nitrogen, methane and carbon dioxide concentrations changes had similar (parallel) trends. Therefore it can be supposed, that after the moment of sealing the area to January 21, 1999 spontaneous heating process didn't develop. Nitrogen was injected from the side of the lower stopping, so it can be supposed, that methane and carbon dioxide moved in the direction of the upper stopping. So behind the upper stopping with a nitrogen concentration increase, methane and carbon dioxide concentrations increased, which wasn't a result of spontaneous heating or combustion of coal. Oxygen concentration decrease corresponds to nitrogen concentration increase. Oxygen concentration decrease mainly results from nitrogen concentration increase. Even though before January 21, 1999 carbon monoxide appeared, no other hydrocarbons appeared, which can be acknowledged as low-temperature spontaneous heating of coal.

In order to confirm the general estimation of the assessed fire status some complex fire indicators were analysed. Trends of complex fire indicators in time are presented in Figures 4.3–4.6. In the case of nitrogen injection into the sealed area, fire indicators basing on nitrogen concentration and oxygen consumption can't be used as a base to interpret the fire status. In this situation only CO/CO_2 index can be used to assess the fire state. The maximum value of CO/CO_2 index behind the lower stopping didn't attain 0.002, and behind the upper stopping it even was below 0.0005 and had decreasing trend in the monitored period. Too small a CO/CO_2 index value indicates that spontaneous combustion didn't occur, only low-temperature heating occurred at the beginning of the monitored period, which confirms the results of the analysis done on the basis of the fire characteristics.

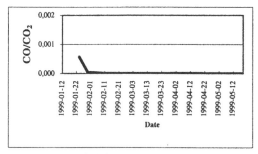

Figure 4.3. Trend of CO/CO_2 index
behind the upper stopping

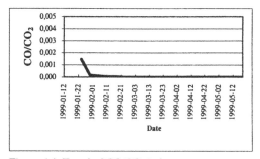

Figure 4.4. Trend of CO/CO_2 index
behind the lower stopping

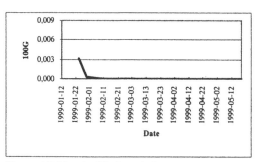

Figure 4.5. Trend of Graham index
behind the upper stopping

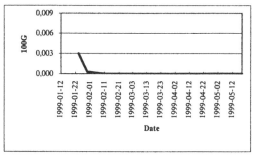

Figure 4.6. Trend of Graham index
behind the lower stopping

CONCLUSION

The suggested in the paper method of sealed-off fire status analysis is based on an analysis of fire indicators time trends and their correlation. The analysis is conducted in two steps. In the first step the fire characteristic is used, which allows drawing preliminary conclusions about the coal spontaneous combustion process in progress, moreover allows to eliminate the factors influencing the results of fire state assessment. The final confirmation the fire status is done by an analysis of compound fire indicators tendencies in time. The suggested method can appropriately indicate the status of a fire in a sealed-off area, which has the vital significance in practice for taking decisions about opening or abandoning the sealed-off fire areas. The cases studied confirmed that the suggested method is beneficial for analysing and assessing the states of fires located in sealed-off areas.

REFERENCES

Brandt S.: Analiza danych. Wydawnictwo Naukowe PWN, Warszawa 1999.
Cygankiewicz J.: Niektóre wskaźniki pożarowe w kopalniach węgla. Biblioteka Szkoły Eksploatacji Podziemnej, Seria z Perlikiem Nr 2, Kraków 2000, ISBN 83-87854-72-7, pp. 45.
Greuer R.E.: Influence of Mine Fires on the Ventilation of Underground. Contract, No. S0122095, United States Bureau of Mines, 1973, pp. 173.
Maciejasz Z., Kruk F.: Pożary podziemne w kopalniach, Część 1, Wydawnictwo Śląsk, Katowice 1977, pp. 11, 22, 24–28, 23.
Mason T.N. i inni: Gob Fires, Part 2, The Revival of Heatings by Inleakage of Air. Paper of Safety in Mines Research Board, No. 76, 1993.
Mitcher D.W.: Mine Fires, Prevention, Detection, Fighting. 3rd ed., Maclean Huster, USA, 1996, pp. 167.
Stark G.W.V.: Use of Jet Engines for Fires Fighting, Tests at the Fire Research Station. Institution of Fire Engineers, Quarterly, No. 60, 1965, pp. 345–360.
Strumiński A.: Zwalczanie pożarów w kopalniach głębinowych. Wydawnictwo Śląsk Sp z o.o., Katowice 1996, ISBN 83-7164-023-4.
Strzemiński J.: Wczesne wykrycie pożarów endogenicznych. Przegląd Górniczy, Nr 5, 1984.
Yuan Shujie: Selection and Estimation of Sealed-off Fire State Indication Methods. Doctorial Dissertation, AGH – University of Science and Technology, Cracow, Poland.

International Mining Forum 2006, Sobczyk & Kicki (eds) © 2006 Taylor & Francis Group, London, ISBN 0415 401178

Development of Strategy for Revival
of Regional Coal-Mining Companies

A.A. Malysheva
Industrial-Ecological Laboratory, The Centre for Monitoring the Socio-Ecological Consequences of Closing the Eastern Donbass Coal Mines Ltd

The established approach to the development of market relations in the TEC (technical and economic complex) of Russia resulted in essential change of fuel and energy balance of the country where in 2004 the share of petroleum and gas made up more than 80%, whereas the share of coal is reduced to a critically dangerous level – 12–14%. At the same time certain mining-geological and investment problems are aggravating and the tendency of the rapidly exhausting natural gas and petroleum prices to rise at the home market is steady. This will inevitably result in the necessity of their partial replacement with coal.

The contemporary coal industry in Russia is very business-orientated, so its main attention of is concentrated on the operation of highly profitable deposits yielding high investment return and located far from the consumers. Besides, carried out without sufficient scientifically-grounded estimation of the available potential, re-structuring of coal industry resulting in closure of more than 200 collieries, essentially aggravated regional social, economic and environmental problems, problems of employment, especially in the areas where coal enterprises were of major importance for the city's economy, that is were city-forming. It should be noted that small coal mining companies commercially unprofitable for objective reasons were much more likely to be doomed for closure than big ones.

The reforms of economic relations in the fuel and energy sector of Russia and the restructuring carried out in the recent years resulted in a number of negative consequences to the functioning of the coal industry: a significant reduction of coal share in the fuel and energy balance (12% – in 2003); a decrease in coal output (60–65% of the level of 1990); lower asset amortization rates (up to 15%); full elimination of small deposits ready for extraction from the stock exchange (due to their objectively low adaptation to the market); under utilization and in some cases complete demolition of the available basic production assets; retrenchment of qualified workers and scientific staff that aggravated social and environmental problems in the regions traditionally focused on coal mining.

A transition to the use of natural gas for power generation and temporarily low internal prices of this commodity connected with an underestimation of the investment profit forming factors, exhaustibility and rarity of its deposits, have considerably lowered the competitiveness of the coal mining industry in Russia as a whole.

Not only individual enterprises but also whole regional coal mining complexes, those representing small and average deposits of local importance in particular, have undergone closure. While the average decrease in coal extraction for the period of re-structuring was less than 50%, some coal-mining regions reduced their coal production ten-fold and more.

Now the production share of coal enterprises working small deposits of local importance makes up less than 1% of the total Russian coal output. Meanwhile, an analysis of the foreign experience shows that countries with available coal resources increase coal's share in their fuel and energy balances (USA, China, Canada), and small deposit operations are encouraged by the state, as for the most part working of small coal deposits promotes development of small independent power engineering, does not do essential harm to the environment and allows to solve problems of unemployment. On average, small coal mining enterprises abroad are responsible for 8 to 15% of the total production.

An intention to increase the coal output and bring the absolute volume of extraction up to 350 million t. by 2020 begins to take shape now, making it necessary to work out a new strategy of coal business development. However, just as it was during the period of re-structuring, insufficient attention of the economists and administrative bodies is devoted today to developing coal-mining companies of local importance, serving the regional market, capable of solving a number of fuel and energy, social, ecological and economic problems by effective utilization of mineral resources available at their territory.

Various methods and aspects of increasing the efficiency of operating coal enterprises in the market conditions have been devised in sufficient detail. Among publications on this subjects the works of A.S. Astahova, N.A. Arhipova, L.A. Belashova, M.M. Gurena, B.A. Davydova, V.T. Kovalja, S.S. Lihtermana, N.J. Lobanova, J.N. Malysheva, A.A. Petrosova, V.P. Ponomareva, V.A. Harchenko, A.B. Janovskogo, M.A. Jastrebinskogo and other scientists are widely known.

An analysis of these works showed that they often disregard modern specificity of development of coal-mining base of local importance and their regional potential, and underestimate the opportunities of their effective utilization for increase of power safety and social and economic stability in the coal mining regions.

From these facts transpires the existence of an actual scientific problem how to perfect the available and develop new methodical approaches to ecological and economic maintenance and effecttive running of regional coal-mining companies, working small and average-sized coal deposits, to promote an integrated use of coal and development of diversification, and the need to solve it. The urgency and the importance of this research for the coal sector and the economy of coal-mining regions will grow in line with the growth of prices of energy resources.

Development of a methodical approach to the maintenance of ecological and economic efficiency of running coal companies working small and average-sized deposits of local importance, promoting growth of extraction, processing and consumption of locally available fuel and energy resources, increase in output of commodity from a unit of extracted and processed resources and employment growth is one of the primary goals of the modern mining economic science.

To attain this aim it is necessary to solve the following problems:

- To analyse the role of local coal resources, associating minerals and waste products of coal production in the development of a regional economy.
- To investigate the condition of regional coal basins after re-structuring and to reveal the basic ecological, social, economic and energy supply regional problems.
- To systematize the internal (regional) and external factors promoting an increase of ecological and economic efficiency of functioning of coal companies.
- To establish the criteria of ecological and economic estimation of the overall performance of coal-mining companies exploiting small and average-sized deposits and using the strategy of diversifications by reclaiming associated useful minerals, waste products of mining and the technogenic deposits formed as a result of partial closure of coal enterprises.
- To work out an algorithm allowing to reveal an ecologically and economically effective way of developing coal companies of local importance promoting growth of coal output with consideration for diversification of production and provision of social living standards of mining workers.
- To develop administrative measures for stimulating the development of small coal businesses and reveal ways to maintain their effective work.

The methodological approach to overcome the above-stated problems utilizes the scientifically proved method of managing the efficiency of working small coal deposits of local importance by increasing competitive commodity output from coal and coal wastes having market value, on the basis of their complex ecological and economic estimation and making use of the specificity of the resource potential of coal-mining regions.

The existing technology of coal production is objectively connected with production of great volumes of solid waste products. Technogenic deposits in places of closed coal enterprises and solid waste products from coal production occupy extensive areas and negatively affect the condition of water and air. The enterprises of the coal sector spend significant means on transportation and disposal of solid waste products, pay environmental faxes for their disposal and the pollution of environment arising from the harmful emissions in places where waste dumps are located.

Transition to chiefly opencast methods of coal mining and paying insufficient attention to an integrated development of coal deposits result in further increase of specific parameters of formation and accumulation of solid waste products per 1 t. of coal, despite the reduction of volumes of produced solid waste products, caused by closing coal-mining enterprises: in 1990 the yield of waste products amounted to 4.2 t, and in 2004 it rose to 7.6 t.

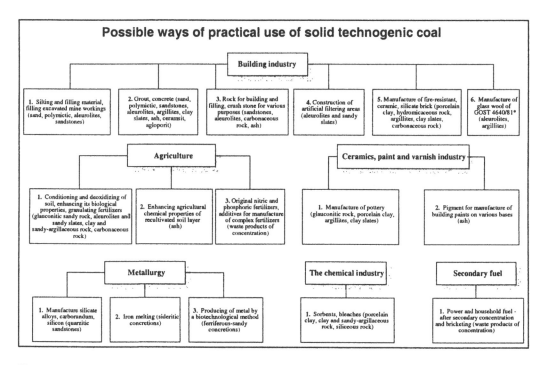

Figure 1. Possible ways of practical use of solid technogenic coal wastes

While the volume of waste products accumulated by the industry is considerable, a mere 3% of them are used as a commodity. Meanwhile, technogenic deposits of coal wastes contain a great deal of useful elements which get an increasing industrial value and social, ecological and economical importance due to the exhaustion and limitation of primary natural resources, lack of financial means for nature protection work and a growing problem of unemployment among miners discharged in the result of closure of coal enterprises. Possible ways of practical use of solid technogenic coal wastes are presented in figure 1.

Thus, the integrated approach to the utilization of small coal deposits of local importance, often developed and ready for extraction, will allow to increase the effectiveness and financial viability of their mining using the existing production assets, qualified manpower and infrastructure and, in conditions shortage of investments in the coal sector, solve the increasing problems of power supply in regions.

Effective and competitive operation of small coal deposits should be fulfilled on a basis of diversification and integrated development of coal resources with the use of market and administrative measures and keeping in mind the need to establish long-term economic relations with the end consumers.

International Mining Forum 2006, Sobczyk & Kicki (eds) © 2006 Taylor & Francis Group, London, ISBN 0415 401178

Mining of Adjacent Coal Seams in the Mines of Western Donbass

Volodymyr Buzylo, Tatiana Morozova, Vitaliy Vasilev, Olexandr Koshka
National Mining University. Dnepropetrovsk, Ukraine

It's well known that mining of adjacent coal seams in the mines of Western Donbass became difficult due to special mining and geological conditions. These stem from the fact of the adjacent coal seams being separated by rock middling between 5 and 20 m, strong and tough coal, having the resistance for cutting of up to 300 kN/m, combined with low mechanical strength of the surrounding rock, rock tendency for separation and swelling, high vulnerability of rock to moisture resulting in 50% up to complete loss of its carrying capacity in humid conditions, disjunctive and frequent disturbances resulting in fractured and weak rock zones.

Mining and geological conditions in all the working mines of Western Donbass are the following: the sequence of preparation is level by level, mining system is the longwall method (from 700 to 1600 m) by coupled or single faces with individual lengths of 150–200 m, mining of seams is carried out in descending order and it takes 2–2,5 years to mine two mining blocks on neighbouring seams with longwalls working in stepped configuration. The mining blocks are developed by raises, all faces are equipped with mechanized longwall sets.

All analytical analyses and calculations of stress and strain conditions of the rock mass available in literature confirm that carrying out stoping in adjacent seams (with rock thickness between the seams of 5–20 m) results in the increase in the intensity of parameters and rock stress in the overmined and undermined zones. Underground investigations were carried out with the aim to prove the truth of the basic theoretical calculations and to obtain actual values of the parameters by which rock stress manifests itself in longwall faces in the situation when adjacent seams are mined.

All investigations were carried out in "Stepnaya" mine OAS MCK "Pavlogradugol" in 2001–2004 in coupled faces 213 and 215, located in the $C_6^{/}$ seam and in coupled faces 117 and 119 in the C_6 seam. The distance between the seams is representative to the area that is the rock thickness between the seams within the perimeters of the examined production sites varied from 8 to 12 m.

Figure 1 shows the mutual positions of the longwall faces in question. The length of the mining block is 1100–1200 m, the length of longwall faces 213, 215 and 119 is 200 m and the length of longwall face 117 is 175 m.

All longwall faces are equipped with KMK 97M mechanized sets with MK-98 powered support and MK 67M cutter loader.

The main goal of the on-site investigations was to establish functional characteristics and dependence of rock stress manifestation parameters in column and in zones of mutual influence of production and development workings.

Investigations were done according to special methodology, taking into account the basic points and demands of universally accepted branch methodology. Actual values of rock joint parameters in the production faces were established. The character of the roof rock and coal seam caving was examined.

While observing stability conditions and rockwall displacement near the face area great attention was paid to fissures, faults, separation of roof rock and cave-ins, the moments of appearing of roof disturbances were established; the distances from the face where they happened were recorded.

The crack width and rock displacements were measured. The character of deformation and rock caving (rock size, thickness of rock fall, caving sequence, and the influence of roof disturbances on the character of roof caving near the face area) was considered. In each case the angle of intersection between the direction of the main jointing of roof rock at the production face and the cleavage present in the coal seam were established.

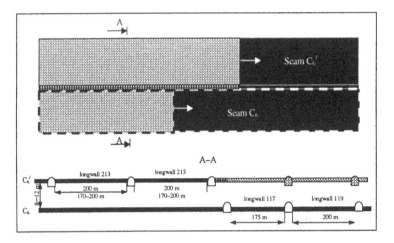

Figure 1. Relative positioning of the examined longwall faces

The results of these analyses showed that the character and regularity of the rock stress manifesting itself in all the examined faces were practically the same. It should be pointed out that in some cases there was up to 0,15 m of coal yield from the seam in faces 213 and 215 and no such phenomenon was observed in the seam C_6. Average collapse volume in the faces in seam C_6' were 15–20% smaller than in seam C_6. Roof rock collapses most often took place in the middle of the faces and at their perimeters.

The fracture formation process in all longwall faces was the same, that is, after passing of the cutter loader vertical cracks appeared, which were opened up having the width of 10–20 m after extracting the next coal seam. After the third cycle the rock of the immediate roof was broken up by chaotically arranged cracks into large pieces of irregular shape, but nevertheless the contour and integrity of block between the technological cracks remained the same. The immediate roof rock caved into the worked-out area after each movement of a powered support section. In some cases a 1,6–3,2 m long cantilever of hanging rock formed above the worked-out area in the faces located in the upper seam.

The parameters of the powered support were recorded with the help of pressure gauge and self-writing manometer MP66A, connected to head ends of powered support hydraulic props. While carrying out investigations concerning support power parameters the pressure of the medium in the hydraulic system was changed in the range of 18–22 MPa being 20 MPa at average. The pressure in the front props was 5–15% higher than in the back ones, which is typical for mining and geological conditions of Western Donbass. As a rule, the support hydraulic props where operating in the condition of increasing resistance. Safety valves of the hydraulic props were setting off in practically all the sections. In the longwall faces in the upper seam the working resistance of the support hydraulic props was in average 5–10% higher, than in faces 117 and 119.

The results of the experiments showed that the actual extraction ratio in all the examined faces was on average 0,95. At the bottom end of the face rock with the thickness of 0,25–0,35 m was mined with the coal to increase the height of the working area because the end head of the main conveyor was situated there and a lot of labour-consuming operations took place there.

Measurement of convergence was accomplished with the help of CYU-II props, installed at benchmark stations in the middle of the face and at its ends. The results of the measurements allowed to establish dependence of roof and foot rock convergence rate from the distance to the technological bench made by the cutter loader's working unit as well as from the distance of 0,1 (Fig. 2) and 1,2 m from the face line.

Measurement analysis showed that the rate of convergence of roof and foot rock at a distance of 0,1 m from the face line begins to increase with the approaching cutter loader as from its distance from the measuring station of 12–10 m and reaches the maximum value at the moment of the working unit passing it. When cutter loader comes back from the measuring station the convergence rate decreases and reaches the background value (0,06–0,08 mm/min) at a distance of 30–35 m. This regularity of convergence rate change is the same when measurement takes place at a distance of 1,2 m from the face.

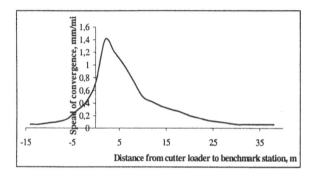

Figure 2. Change of convergence rate of roof and foot rock depending on the distance to the cutter loader

Figure 3 shows the curves of changing total convergence value of rock in coal seam in dependence on the distance to the cutter loader's working unit in 117 (curves 1 and 2) and 213 (curve 3) face.

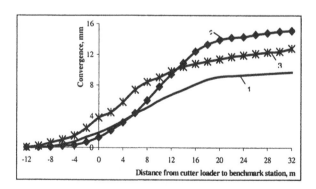

Figure 3. Dependence of longwall convergence value on the distance from the cutter loader.
1 – benchmark station at a distance of 0,1 m from the face line in longwall face 117;
2 – benchmark station at a distance of 1,2 m from the face line in longwall face 117;
3 – benchmark station at a distance of 0,1 m from the face line in longwall face 213

It was established that the change of value of the total convergence of roof and foot rock in the face is subjected to the same laws in the case of mining the seam in pillars (face 213, curve 3) as in face 117 (curve 1). Both curves are described as regression equations of the second degree with satisfactory convergency:

- the 1st curve $y = -0,006\ x^2 + 0,45\ x + 4,57;\ R^2 = 0,98;$
- the 3rd curve $y = -0,003\ x^2 + 0,32\ x + 2,83;\ R^2 = 0,96.$

The comparison of measurement results shows that the total value of rock convergence near face area (benchmark station at a distance of 1,2 m from the face line, curve 2, Fig. 3) is approximately 30% more than in the face (bench mark station at a distance of 0,1 m from the face line, Fig. 3, curve 1). This difference is explained by increasing distance between the face and measurement station after passing of the cutter loader and another reason is that support doesn't exert influence on roof rock.

Measurements of rate and value of convergence were not conducted in the measuring station situated at the end of the face. It was established that the rate and general value of convergence in the end of the face is 15–20% higher than in the middle of the face. The reason for this phenomenon are the zones of high stress situated at the end of the face, created by interacting bearing stress zones from the face and development workings.

The research into the value and rate of roof and foot rock convergence in the face allowed establishing the following regularities:

- Zones of intensive convergence are formed not only due to coal winning but also due to regular unloading of powered support in the state of its movement.
- The cutter loader speed greatly influences the total value and rate of convergence of the enclosing rock, that is, the higher the speed of the cutter loader, the higher the convergence values and rate.

For example, convergence rate outside the zone of cutter loader operation (we call it background rate) was 0,04–0,06 mm/min when its feed rate was 1,8–2,5 m/min and 0,06–0,08 mm/min when the feed rate was 3,3–3,5 m/min:

- In some cases there was an abrupt increase of convergence value and rate in front of technological bench created by the working unit of the cutter loader while extracting the seam, which is explained by the bench crushing due to high stress concentration within the solid coal seam.
- Convergence rate decreases when cutter loader approaches a development working, which is confirmed by the presence of a zone of crushed and squeezed coal at the intersection between the longwall face and the gate.

An analysis of the results of rock convergence in the working area of a longwall face allowed coming to the following conclusions:

- The process of rock displacement at the end of a face is characterized by higher intensity than in the middle of a production face, so while developing a technological layout of extracting seams in special mining and geological conditions it is required to plan for a set of measures directed at strengthening the the face - development working intersection area.
- Convergence is greatly increased in the case of increasing feed rate of the cutter loader, so on the unit approaching the gate its feed rate must not be greater than 1,5–2,0 m/min as from a distance of not less than 8 m.
- To decrease rock convergence it is necessary to move the end section of powered support and conveyor heads at maximum cutter loader distance from the intersection.

According to the results of the underground experiments it was established that the character and regularity of rock stress manifestation while mining out upper and lower seams are practically the same, only the values of the parameters are different. By the way, these values are higher in faces operating in pillars, than in faces operating in overmined conditions. This is explained by the fact that rock stress after mining out the top seam and before starting to extract the bottom seam (2–2,5 year) stabilizes, the rock sag, however, continues.

International Mining Forum 2006, Sobczyk & Kicki (eds) © 2006 Taylor & Francis Group, London, ISBN 0415 401178

Areas of Sustainable Development in of Mine Closure Process

Jan Palarski
Technical University of Silesia. Gliwice, Poland

ABSTRACT: Mining industry is the important contributor to the Polish economy. The mineral sector creates many opportunities including: jobs, development of infrastructure and services in mining regions, transfer of technology and research results. However, the operational live of mines is finite. Due to the exhaustion of reserves and high operational costs, mines need to be closed.

This paper considers many aspects of mine closure policy process. Mine closure plans should include site closure problems as well as economic, environmental, social and employee matters. The paper provides information on technical, social, ecological and economic aspects of mine closure in Poland. Some of these are related to physical stability of rock mass, water management and chemical stability, tailings and waste disposal facilities, post closure land use and environmental monitoring.

KEYWORDS: Sustainable development, mine closure, surface damages, filling and grouting process

INTRODUCTION

Over the last fifteen years Polish mining industry has been put under tremendous pressure to improve its social, economic and environmental performance.

Since 1989 Polish mining sector has closed 32 coal and 9 lead and zinc, cooper, barite and salt underground mines. In addition, many sand and limestone quarries and borehole mines (salt, sulphur, oil and gas) were closed in this period.

In years 2003–2004 total or partial closure processes have been carried out in following mines:
- 7 underground hard coal mines.
- 2 underground lead and zinc mines.
- 2 underground and 2 borehole salt mines.
- 1 underground coppers mine.
- 15 oil and gas mines.
- Ca. 250 open cast mines of different minerals.

Mining companies decided to undertake new programs of internal reform, which aim at achieving a serious change in the mineral sector. This initiative in coal mining industry was called "Polish Coal Mining Restructuring Program". In order to improve the competitiveness of the sector, the program focused on people, technology process and environment. The key elements of the restructuring program were changes to existing legislation, policies and working practices. The objective was to convert the mining industry into a fully commercial business. To achieve this, significant increase of productivity and the cost reductions were needed.

The program included a number of elements:
- Restructuring of management process.
- Closing of inefficient or uneconomical mines.
- Reduction of workforce and production capacity.
- Continuing advances in underground mining technology.
- Protection of environment and public health and safety.

Coal plays a significant economic role in Poland. Hard coal produced in country is mainly from underground mines in the Upper Silesian Basin (96%). Coal from this region is produced by three state controlled companies that operate 36 mines and by two independent mines (Table 1).

Table 1. Longwall specification (2005)

	Factor	
1.	Number of longwalls	125
2.	Mining depth (average)	724 m
3.	Panel width (average)	218 m
4.	Panel length (average)	860 m
5.	Cutting height (average)	2.35 m
6.	Number of gate entries	2
7.	Depth of cut	0.8 m
8.	Average production	2909 tonnes per day

Around 20 Mt of hard coal production is exported to Europe. Approximately 80% (about 80 Mt) of total hard coal production is currently utilized by power stations (electricity and heat production), steel and cement industry and for domestic consumption. Coal is used to generate 96% of total Polish electricity. The current mining practice in Poland is dominated by longwalls operating in seams from 1.5 m to 4.5 m high at depth from 300 m to 1200 m.

Currently, one of the most important aspects of Polish coal mining industry is a proper preparation of mine closure plan and its implementation. The paper will provide information on technical, social, ecological and economic aspects of mine closure according to the principles of sustainable development.

SUSTAINABLE DEVELOPMENT FOR THE MINING

The widely accepted definition of sustainable development is the one used by the World Commission on Environment and Development (1987): "Sustainable development is development that meets the needs of the present without compromising the ability of future generations to meet their own needs". Mining has a key role to play in assisting the mineral sector to make an essential contribution to sustainable development of mankind. Sustainable development means on one hand stability and on the other hand change. Mining needs to demonstrate continuous improvement of its economic, social and environmental contribution to sustainable development. The stability refers to the fundamental processes of the economy, environment and culture, but change means necessity of permanent innovation. It is often questioned if mining can be sustainable, because it extracts finite resources. Thus, mining reduces the potential for future generations and it does not contribute to sustainable development. This is a narrow interpretation of the ideas of sustainability. Mining contributes in a wider sense to the particular principles of sustainable development. First of all, mining should maximize economic, ecological, and socio-cultural benefits from mineral extraction and contribute to the improvement of local development.

One of the most important aspects of mining management is the proper preparation of mine closure plan and its implementation. Generally closure planning should start during the pre-feasibility phase of a mining project, and should be supplemented and improved during mine life to ensure a successful final closure. Every closure plan considers the long-term social effects and physical, chemical and biological impacts on the ecosystem.

Mine closure is much more than cessation of mineral production and removal of infrastructure – decommissioning. The mine closure also includes rehabilitation of site, which means the return of disturbed land to a stable, productive and self-sustaining condition. In addition, the socio-economic issues of closure and the impact on mining communities and the local economic development should be addressed. Not long ago when mines stopped operating, they were simply abandoned and flooded. The traditional approach towards underground mine closure focused on surface protection, sealing and/or filling shafts, adits, underground workings and where needed, location of pump stations and monitoring systems. Today, responsible mine closure additionally involves removing unwanted equipment and infrastructure, securing waste dumps and impoundments, protecting ground water, neutralization of hazardous wastes and rehabilitation of land (Figure 1).

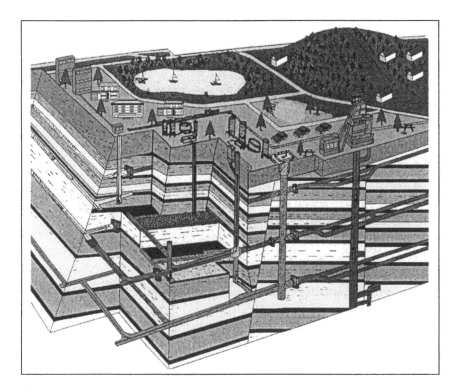

Figure 1. Scheme of a closed mine

Polish coal mines are being closed on the basis of the closure plan elaborated earlier and according to the program of solving the socio-economic problems. The plan includes the rules of safe closing of the underground workings as well as demolition or restoration of surface infrastructures. It also includes the reports on gas and water hazards, on mining damages and the information on after use of the site. The rehabilitation plans and monitoring systems prepared as part of the environmental impact assessment are tested and verified.

MINE CLOSURE

During the restructuring process of Polish coal mines, in the period of 1989–2004, 24 collieries were closed and 16 others were combined into 8 mines. The location of both operating and closed mines is presented in Figure 2. It shows clearly that the closed mines are situated in the Northern part of Silesia Coal Basin and in Lower Silesia Coal Basin. In those regions it was necessary to fill or seal about 3000 km of underground workings and 92 km of shafts. In addition 2300 buildings and many surface structures were demolished. The mining damages were removed in about 4000 buildings and infrastructure and over 1100 hectares of mined-out area was rehabilitated.

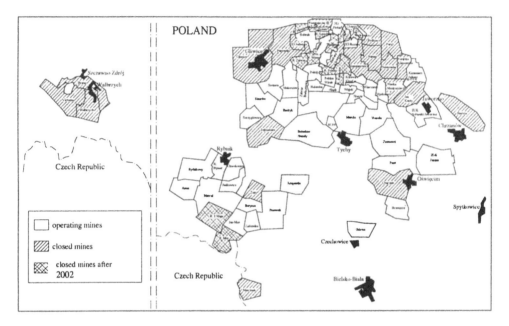

Figure 2. Operating and closed mines in Poland

Since 1993 the number of miners working in the coal mines has declined from about 312,000 to about 128,000 in 2004.

Mining in each abandoned colliery in Silesia was carried out in at least several seams. A dozens of mines has operated at the depth below 20 m. In addition, in this region at the depth of less than 300m underground mining of zinc and lead ores was carried out on the area of the coal mines. All the collieries were operating under the built up areas, including urban districts, industries, roads, motorways, railways, pipelines, rivers, etc. Apart from that methane occurs, the coal has a tendency to spontaneous combustion and still huge amounts of water – usually over 5000 m³/day/mine – are running there (Figure 3).

Those complicated geological conditions, existing infrastructure on the surface and vast areas of mining activity require the use of special mine closure methods, removal of mining damages, water pumping as well as monitoring. When the mine is abandoned and closed, the greatest attention is paid to the following issues:

– Minimisation of the surface damages and creation of the depressions and sinkholes over abandoned workings.
– Removal of all unwanted plants and equipments.

- Secure waste impoundments and dumps.
- Use of proper methods for inset protection, filling the shafts and construction of clay seals or plugs during the general shafts filling.
- Use of proper mine ventilation during the removal of unwanted infrastructure, sealing and filling operations.
- Designing a mine water scheme, which will totally satisfy the requirements of water capture, pumping, treatment and effluent disposal or detoxifying.
- Site restoration and monitoring.

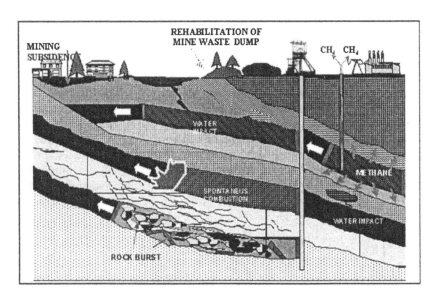

Figure 3. Hazards in closed mines

ABANDONED MINES INFLUENCE ONTO THE SURFACE

Old and shallow workings pose great danger for the buildings, infrastructure and the people. Over the workings of up to 100 m deep, there are sinkholes and depressions created on the surface. However, there are some large surface subsidence depressions coming from workings deeper than the above ones. To eliminate the surface effects and to minimize the danger of gas outlet and underground fire, all shallow workings in old mines should be filled with backfill materials. Not all the roadways can be reached from other underground workings, so there is a need to do the boreholes from the surface. They are situated in such a way the best filling of the voids can be achieved and sudden inrush of water can be avoided, otherwise, cavings around the bore-holes can be expected. If possible, before filling the void an exact estimation of its volume should be done by means of TV camera put through the bore-hole down on the line. The camera survey can determine areas of roof fall and the distance from the hole to various objects as mine entries, coal pillars, voids and fractures.

Void filling is done through the bore-holes of 120–200 mm diameter, using the following methods:
- Gravity placement method of dry waste material (crushed rock, surface sand, gravel, slag and ashes) – usually the material is transported through a well from surface directly into a void.
- Pneumatic transport of the dry fly-ash with the capacity up to 30t/h, while the pressure is about 0,25 MPa.

- Pneumatic transport of the fly-ash mixed with water at the outlet of the quantity of up to 20 l/min.
- Supply by gravity or by means of the pump of the ash-water mixture or mixture of ash (up to 70% of the dry mass), tailing (up to 25%), cement (up to 10%), and calcium chloride (2%) – concentration of the mixture is to be decided each time according to the state and conditions of the void filling (mixture concentration < 75% by mass).

The last solution is most effective and is most often used in Polish collieries. The slurry of fill material is gravity fed or is pumped down through a well into the underground voids until the well will not accept any additional mixture. Slurry injection under pressure allows gaining increased distribution of the fill material within mine workings and caving areas. Properly chosen quality and quantity of fill mixture, appropriate borehole spacing, and injection pressure allow achieving high fill factor and adequate distribution of the fill mixture within mine workings and caving areas. By means of geophysical techniques the fill factor of underground voids can be controlled. After the fill process is stopped, it is important to ensure that the wells don't have a pathway. All fill wells should be plugged and sealed with grouting mixture or cement from the bottom to the surface.

During backfilling, many samples of mixture are collected for testing of slump, viscosity, compressive strength, elastic modulus, hydraulic conductivity, swelling strain and chemical parameters (pH and trace elements leached from fill mixture).

The best results are achieved by filling of voids and tunnels or by grouting of caving areas directly from underground workings. In this case a broad assortment of mixture can be used for filling and grouting processes. These mixtures may include coal combustion by-product, tailing, sludge, sand and binder. For filling operation a special distribution plant with gravitational transport is used (Figure 4).

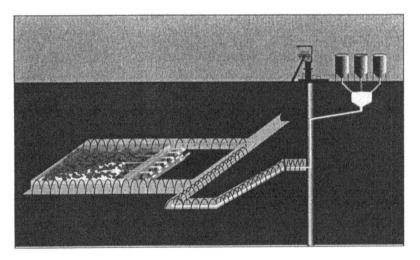

Figure 4. Scheme of a grouting plant with injection pipe drawn behind support

Figure 5 presents an example of the bore-holes location along the railways running over the workings driven in years 1870–1880. Up to 5 m high voids are situated at the depth of 47 m up to 74 m. To protect the railways, viaduct and a part of the highway, the fill of the workings on 16,000 m^2 area had to be done. During the period of 14 months about 23,000 t of the fly ash-cement-water mixture of concentration 69.5% by mass was gravitationally supplied into the underground voids. It fully protected the surface in this region. The experiments proved that to fill 1 m^3 of void about 1.3–1.8 m^3 of the mixture depending on the cracks and the concentration of slurry is needed.

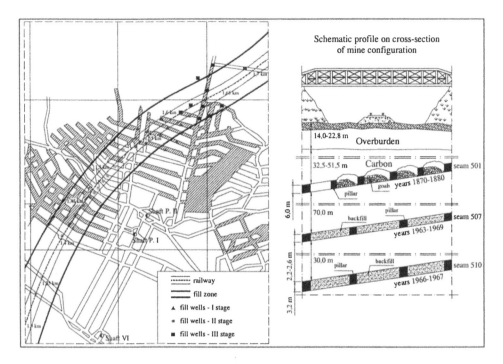

Figure 5. Example of the bore-holes location along the railways running over the workings driven in years 1870–1880

SHAFT FILLING

The abandoned shafts are filled or they are left without any filling, for example to be used for pumping water. Before filling the shaft all the equipment, guides, cables, ropes, pipes or ladders must be removed. Shaft insets must be prepared for constructing the barricades, fill dams or plugs (Figure 6). Hardcore in the form of broken rock is recommended for shaft inset locations and sumps. The remaining part of the shaft can be filled with mining waste, coal combustion by-products as well as with demolition materials (crashed brick, concrete). When water or gas flows into the shaft it is necessary to prepare clay seals; their thickness and depth depend on geological conditions near the shaft, technical state of its support, material used for its filling and seal construction. Table 2 presents basic requirements for the shaft filling materials and construction of the seals.

Table 2. Properties of fill and construction material

	Factor	Unit	
		Shaft fill material	
1.	Fill material	–	General fill: mine waste, demolition material, limestone Hardcore fill: dolomite, basalt at inset and in acidic conditions Non-toxic, non-combustions
2.	Quartz content	%	25
3.	Calorific value	kJ/kg	≤ 16.000
4.	Settlement of fill	%	≤ 7

137

Table 2 cont'd

	Factor	Unit	
		Seal (plug)	
1.	Construction material	–	Clay, bentonite, concrete
2.	Permability	m/s	$\leq 10^{-9}$
3.	Plasticity index	%	10–20
4.	Lenght of clay seal	m	5–50 (average 15 m)
		Dam	
1.	Construction material	–	Concrete, brick, timber sand, fly ash, roof fall rock
2.	Length of dam	m	0,5 · diameter, minimum 5 m
3.	Strength of grouting rock mass at the interface of seal or dam	MPa	≥ 10

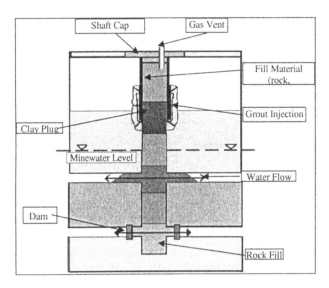

Figure 6. Scheme of a shaft filling

Capping represents the final operation of mine closure. The shafts are protected by the rein-forced – concrete plate of the diameter that is twice as big as the diameter of the shaft. There is a 600 mm diameter concrete or steel pipe for later filling and a gas vent in the shaft capping.

PUMPING AND DISPOSAL OF MINE WATER

Almost all the abandoned and operating mines in Silesia are connected either by workings, cracks or fractures in boundary pillars and barriers. Accumulated mine water in abandoned collieries could cause an unacceptable risk to adjacent operating mines due to high static pressures developing on barriers and the risk of breaking through strata. Operating mines are protected by various strategi-cally located pumping stations.

To simplify the drainage system and minimize the risk the abandoned mines were grouped into 4 systems regarding the hydraulic connections among them and protecting mines in operation as well as reducing the costs of pumping (Figure 7).

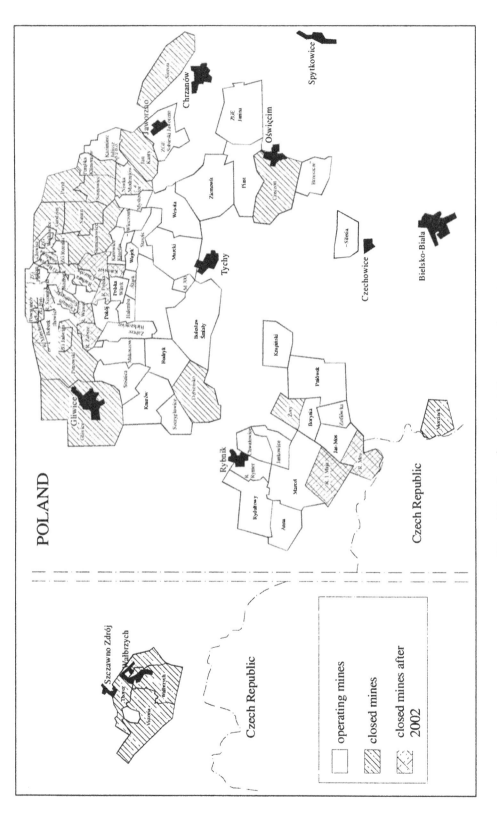

Figure 7. Location of pumping stations – pumping systems

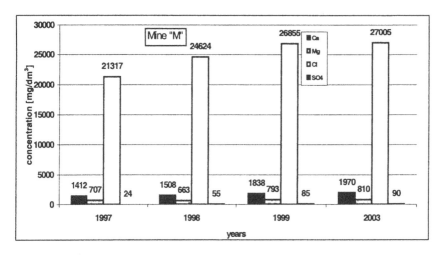

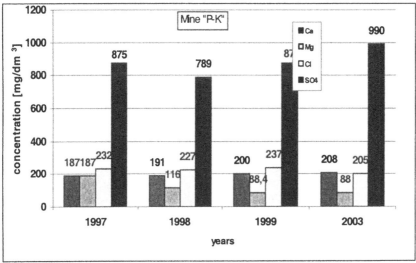

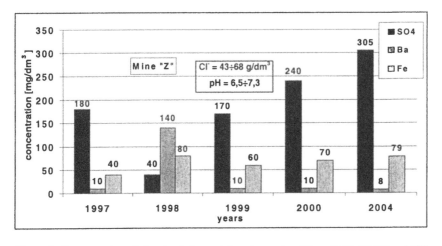

Figure 8. Concentration of dissolved elements in mine water, Mines "M", "P-K" and "Z"

The quality of mine waters depends on chemical processes that take place in the strata, gob area and workings especially in those filled with the fly-ash and in the strata containing pyrite. The products of pyrite oxidation that dissolve in water cause the increase of SO_2 ions and the appearance of acid reaction. In general, in Polish collieries there are mine waters containing huge amount of sodium chloride (salt) up to 130 g/l. After mine closure, there is a visible tendency of chlorides' decrease in the mine waters that may be a result of great water inflow from the overburden to carbon strata through cracks and workings where the partial salt washing had already taken place. In regions where backfilled workings with fly-ash become flooded after mine closure, there is an increase of water alkalinity (pH rise), sulphate penetration into the saline waters and a decrease of dangerous barium ions that are precipitated in the waters containing sulphates and remained in the rock mass (Figure 8).

CONCLUSIONS

– Polish coal mining industry faces a number of challenges. To address the economic challenges, a major reorganisation of industry takes place and it aims at full commercialisation.
– First of all, mining should maximize economic, ecological, and socio-cultural benefits from mineral extraction and contribute to the improvement of local development.
– Mining restructuring caused non-profitable mines closure and introduction of the more effective mining technologies that are useful for the environment.
– Closed mines cause a huge danger for the surface when improper fill methods of abandoned workings and shafts are introduced. Moreover, special methods of dewatering of abandoned mines must be introduced to prevent working mines from flooding. In Poland, for the closure of coal mines, the ways of safe filling of the workings and shafts have been worked out as well as the systems of central dewatering. At the same time continuous monitoring of the water quantity and quality and mining damages is being carried out.

REFERENCES

Coal Statistic. Facts about Polish Coal Mining 2003. Katowice 2004.
Palarski J.: Recent Developments and Trends in Filling of Polish Mine Workings. Proc. of 10th International Conference on Transport and Sedimentation of Solid Particles. Wroclaw 2000.
MMSD. Breaking New Ground. Mining, Minerals and Sustainable Development. IIED. Earthscan. Publications Ltd, London and Sterling, VA. 2002.

International Mining Forum 2006, Sobczyk & Kicki (eds) © 2006 Taylor & Francis Group, London, ISBN 0415 401178

Potential for Improvement of Reliability and Maintenance in Mining Operations Based on Nuclear Industry Know-How and Experience

D. Komljenovic
Hydro-Québec. Gentilly, PQ, Canada

J. Paraszczak
*Dept. of Mining, Metallurgical and Materials Engineering, Université Laval.
Quebec City, PQ, Canada*

V. Kecojevic
College of Earth and Mineral Sciences, Pennsylvania State University. University Park, PA, USA

ABSTRACT: This paper discusses possibilities concerning a transfer of know-how and technology transfer in the field of reliability and maintenance from the nuclear industry to mining. Both industries face important challenges with regard to safety, productivity and environment. The nuclear industry has achieved significant levels of productivity while remaining safe through a systematic approach using accumulated knowledge and developing industry tailored Reliability and Maintenance Processes. Some examples of their potential application in mining have been given. The paper presents also some indications and suggestions concerning further research work in this field.

1. INTRODUCTION

Production and cost figures for mining operations depend considerably on performance of the equipment employed. As working conditions, particularly in underground mines, are extremely demanding and harsh, and most of production systems have little or no redundancy equipment failures may severely disrupt production, causing substantial losses. At the same time, direct maintenance costs in open-pit and underground mines in North America, Australia and Chile often account for over 30% of the total production cost [Campbell 1995, Knights 1999, Lewis 2000, Knights & Oyaneder 2005]. This is partly due to the fact that at many mine sites, particularly the underground ones, a substantial part of maintenance is still reactive; sometimes well over 50% [Paraszczak *et al.* 1997, Knights & Oyaneder 2005]. This is very far from 15% and 20% benchmarks given by Mitchell (2002) and Caterpillar (2002) respectively. In these circumstances, equipment reliability as well as maintenance actions intended to prevent the occurrence of failures are among the key concerns for mine operators. Unfortunately, it seems that scheduling and scope of proactive (preventive and predictive) maintenance tasks recommended and required by Original Equipment Manufacturers (OEM) cannot be considered optimal. Manufacturers' generic maintenance programs do not take into account substantial differences between various mine sites in terms of requirements and operating and service conditions. Due to that, recommended maintenance tasks are often too cumbersome, unnecessarily time-consuming and costly, up to the point where their scope, content, frequency or even pertinence may be questioned. There is also a potential conflict of interest for equipment manufacturers between developing a maintenance program optimal for a customer and the one that is the best for them. Therefore, in order to enhance equipment reliability and availability and, consequently, to reduce production cost, it is desirable that maintenance procedures and policies be optimised. With this in mind, some companies have attempted to implement methodologies and procedures already "tested" in other industries, such as for example Reliability Centered Maintenance, known under its acronym "RCM". Accordingly to Ednie (2002), RCM has been used in Canada by Iron Ore of Canada, Quebec Cartier Mining (iron ore mines), Syncrude's Aurora tar sands mine and BHP Ekati diamonds mine. Although in general terms the improvement of equipment reliability and a decrease of maintenance costs have been reported, little or no detailed information about the actual results has been provided. RCM practices were to be implemented also by Falconbridge's Raglan underground mine [Mercier *et al.* 2004]. As the opinions on practicability of RCM for the mining industry are far from unanimity, it is worthwhile to consider other potentially beneficial "know-how/technology" transfers. In this context, this paper presents an overview of Risk-Informed Asset Management (RIAM), AP-913 Equipment Reliability and Preventive Maintenance Optimisation processes (ERP nad PMO, respectively) developed for the nuclear industry. It discusses also an issue how the experience and lessons learned there might be useful and beneficial in terms of improvement of mining equipment effectiveness, environment protection and safety.

2. RELIABILITY AND MAINTENANCE – APPROACH AND EXPERIENCE OF THE NUCLEAR INDUSTRY

Despite substantial differences in the nature of the mining and nuclear industries, some analogies between them may be found. Nuclear power stations are also expected to produce more (energy in its case) at a lesser cost, while respecting increasingly demanding safety and environmental standards and requirements. Similarly as in the mining industry, the optimisation of equipment reliability and maintenance are among the key contribution factors with regard to achieving these objectives. Following the efforts of World Association of Nuclear Operators (WANO), its Institute of Nuclear Power Operations (INPO), regulatory bodies, and numerous research institutions (in particular Electrical Power Research Institute – EPRI), the nuclear industry has developed its own reliability and maintenance processes. An integral Equipment Reliability Process (ERP) AP-913, and Preventive Maintenance Optimization (PMO) have been successfully introduced throughout the industry.

These two processes (i.e. ERP and PMO) are an integral part of a more global process developed in the nuclear industry named Risk-Informed Asset Management (RIAM). The latter will be discussed more in detail in the next section. It should be stressed that ERP and PMO may be implemented as standalone at a plant (or at another complex facility) without implementing the whole RIAM process.

2.1. Risk Informed Asset Management (RIAM) for Nuclear Generating Plants

The economically regulated nuclear generating industry has developed several physical and financial asset management tools. Life Cycle Management (LCM) and business planning are being done at almost all commercial nuclear power plants. Formerly, LCM was focused on reliability improvement and cost reduction. Nowadays, the evaluation of plant improvement projects is generally based on best-estimate, point-value technical (safety and reliability) and economic (net present value – NPV) methods [Liming 2002].

In this context, the industry has been developing a process for risk-informed asset management (RIAM) of its facilities. Its main objective is to develop a rigorous systematic risk-informed approach for assessment, analysis, prediction, and monitoring of power plant economic (i.e. financial) performance while maintaining high confidence levels that pre-established safety limits are not and will not be breached [DOE-DP-STD-3023-98 1998, NUREG-1513 2001, Liming 2002, NUREG-0800 2002, Coppock 2004, Messaoudi 2005].

The RIAM process consists of modeling and probabilistic quantification of decision support performance indicators. It assists decision-makers in determining not only which facility improvement investment options should be implemented, but also how to prioritize resources for their implementation based on their predicted levels of profitability [Liming, 2002]. Key decision support indicators include, among the others:

- Net Present Value (NPV).
- Projected Earnings.
- Projected Costs.
- Nuclear Safety (core damage frequency, large early release frequency, etc.).
- Power Production (availability, capacity factor, etc.).
- Efficiency (heat rate).
- Regulatory Compliance.

The RIAM approach complements and integrates in-plant existing activities such as probabilistic risk assessment (PRA), equipment and component reliability process (ERP), preventive maintenance optimisation (PMO), and life cycle management (LCM) methodologies. RIAM involves the integrated assessment of many characteristics and performance measures related to nuclear power generating stations. This process is intended to maximize both net present value (NPV) of the facility, and long-term profitability through a continuous support to a decision-making process.

RIAM introduces numerous models and supporting performance metrics that can ultimately be employed in order to support decisions that affect the allocation and management of plant resources (i.e., financial support, employment, scheduling, etc.). The experience has also shown that decision support metrics should be applicable to sets of multiple improvement options in order to enhance effective resource optimisation.

While the initial applications of RIAM have been developed for nuclear power stations, it can be adapted to provide a decision-making support to other types of power stations, complex facilities (usually capital-intensive), or even groups of such facilities across a wide variety of industries. Figure 1 presents a general block model of the RIAM process.

This paper focuses mostly on Equipment Reliability Process (ERP) AP-913, and Preventive Maintenance Optimisation (PMO) as integral parts of RIAM process. These processes are incorporated more or less into following models: safety, generation, availability, cost, and efficiency.

2.2. Equipment Reliability Process (ERP) AP-913

Institute of Nuclear Power Operations (INPO) has developed a fundamental equipment reliability process AP-913, aimed to assist member utilities to maintain high levels of safe and reliable plant operations in an efficient manner (INPO, 2001). INPO defines ERP as an integration and coordination of a broad range of equipment reliability activities into one process. It is developed for the plant personnel to evaluate important station equipment, conceive, work-out and implement long-term equipment health plans, monitor equipment performance and condition, and make continuing adjustments to preventive maintenance tasks and frequencies based on equipment operating experience (INPO, 2001). This process includes activities normally associated with such programs as classical RCM, or with some of its streamlined versions, preventive maintenance (periodic, predictive, and planned), Maintenance Rule, surveillance and testing, life-cycle management (LCM) planning, and equipment performance and condition monitoring.

INPO defines the following objectives for this ERP:
- The process should be efficient, incorporate human factor considerations, and ensure effective performance during all phases of plant operations.

- In the case of a company operating several plants, a uniform process to be used by all of them.
- The lessons learnt in-house and throughout the industries that are applicable are to be incorporated into the process to improve adequacy and efficiency.
- Changes to the process should be timely, responsive to user's feedback, and implemented at all affected plants.

ERP also lists various equipment performance objectives. The top-level diagram of this process is shown in Figure 2 [INPO 2001, Komljenovic *et al.* 2004]. It includes major process elements and their relationship. The reference AP-913 [INPO 2001] also gives a process flowchart including detailed activities indispensable to achieve the process objectives.

The intent of AP-913 is to identify, organize, and integrate equipment reliability activities into a single efficient and effective process. When available, RIAM, with its treatment of risk and plant-level life-cycle equipment reliability, will be a tool to assist utilities in implementing AP-913 process (directly incorporated into both safety and generation models, see Figure 1).

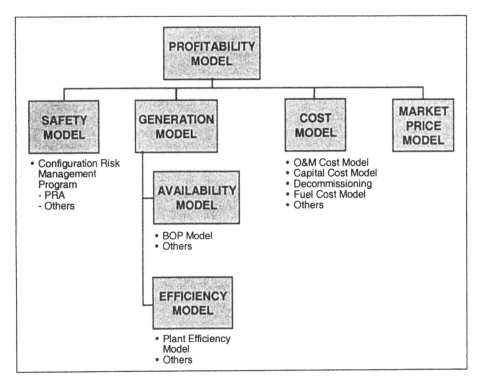

Figure 1. RIAM Conceptual Model Outline [Liming 2002]

The potential benefits stemming from the application of the integral AP-913 ERP are as follows:
- Consistent, systematic and rigorous approach to major investment value-based decision-making.
- Consistent treatment and integration of plant safety, reliability, efficiency, and cost factors in the decision-making process.
- Consistent framework for continuous monitoring, and projection of plant safety and production performance.
- Operating & Maintenance procedure improvement and procedure training prioritisation.

- – Trade-offs between on-line and off-line maintenance.
- – Capital spares procurement analysis and prioritisation.
- – Unit efficiency improvement, etc.

2.3. *Preventive Maintenance Optimisation (PMO)*

Following several references, a standard or classical RCM was not found to be an optimal process with regard to the needs and expectations of the nuclear industry. It usually led to the creation of a new preventive maintenance program rather than to enhance and revise the existing ones. Furthermore, a considerable amount of time and effort required to perform an analysis and implement PM tasks recommendations has been considered as another major weakness for standard RCM applications [Johnson 1998, Turner 2002, Harazim & Ferguson 2003, Messaoudi *et al.* 2003, Messaoudi 2005].

The experience has shown that approximately 60% of the time required to accomplishing a classical system RCM analysis is spent on the following two major steps:
1. Failure Modes and Effects Analysis (FMEA) – identification of equipment critical for each system function.
2. Logic Tree Analysis (LTA) – identification of the most feasible and effective PM tasks to prevent critical component failure modes and causes of concern.

This situation is due to the level of details required and excessive documentation produced as a result of the rigid process steps. In this context, in mid-1990s the nuclear industry initiated research projects to develop ways to perform RCM more cost effectively, based on its own experience. Some concepts have been developed in order to demonstrate that the conventional RCM methodology [Moubray 1997] could be streamlined to reduce the cost of analysis while maintaining a high quality product. More information about the results of these initiatives (works of EPRI, Fractal-solutions, INPO, etc.) may be found, among other sources, in the work of Messaoudi (2005).

The PMO concept is based on a more streamlined RCM approach. With PMO, many nuclear generating stations have achieved very high capacity factors (over 90%) at a lower cost, while ensuring a safe production [Johnson 1998 and 2003, Turner 2002, Harazim & Ferguson 2003, Messaoudi *et al.* 2003, Coppock 2004, Messaoudi 2005]. Table 1 (below) shows a comparison between the classical RCM analysis and streamlined methods.

Table 1: Comparison of various RCM related concepts after [Messaoudi 2005]

| Classical RCM analysis phases | Streamlined analyses | | |
	Streamlined Classical Process	PMO Process	Criticality Checklist
Determine System Boundaries	X	X	X
Determine Subsystems	6)		
Data Collection and Plant History Review	X	X	X
Identify system Functions and Functional Failures	1)	1)	
FMEA	2)	3)	4)
Instrument Matrix[7]	X		
Non-critical Evaluation	X	X	X
LTA PM Task Recommendations (Critical/Non-critical)	5)	5)	5)
Task Comparison	X	X	X

147

1) Identify all system functions, and classify into two major groups: (1) Important functions, and (2) Non-Important functions.
2) Perform FMEA only on components and equipment that support important/critical functions.
3) Perform a streamlined FMEA, which combines dominant failure modes and plant effects into one component record on components (equipment) that support important functions.
4) Component (Equipment) is critical if its failure will result in one of several pre-selected plant effects, which one wants to prevent through a PM program.
5) Maintenance Templates added to formal LTA. LTA only formally documented in modified classical analysis.
6) Only for very large (complex) systems.
7) Replaces FMEA for Instrumentation & Control (I&C) equipment.

As shown in Table 1, the PMO employs many of the same analysis techniques as the classical RCM, but it represents a more streamlined approach. The classical RCM analysis begins at the top with a system, breaks it down into subsystems, identifies critical components, and recommends PM tasks. The latter are subsequently compared to those already in place (existing PM tasks), which leads to final task recommendations. Contrary to that, PMO begins at the opposite end. The PM procedure is split into tasks that are reviewed to identify the failure they are intended to prevent. Subsequently, related data are collected, critically reviewed and analysed in order to make final task recommendation [Johnson 1998, Messaoudi 2005]. Another major difference in the PMO process vs. the classical RMC analysis is that the standard FMEA and LTA have been streamlined and combined into one record (see Table 1). The PMO process documents the most important failure modes for a component/equipment (usually three dominant ones), and allows documenting the most dominant plant effects (typically three) in one component/equipment record, used to determine component/equipment degree of criticality. For those considered critical, PM tasks are recommended using maintenance templates.

The most significant elements contributing to the successful completion of the system analysis using this concept are as follows:
– Combination of failure modes and plant effects into one record.
– Combination of FMEA and LTA into one component record with less formal documentation.

Efficiency and savings related to performing LTA in the nuclear industry have been achieved through a concept of "Maintenance Template" for a given component or equipment type. At the stage of a template development, several component characteristics such as equipment frequency of usage, operational and service environment as well as functional importance are taken into account. Each maintenance template recommends appropriate tasks for a given component or equipment type in terms of condition monitoring tasks, time directed tasks, and surveillance tasks. They provide major additional savings. In the nuclear industry there currently exist maintenance templates for nearly 80 major components and/or equipment considered critical with regard to both safety and power generation. They have been developed through a teamwork involving manufacturers, nuclear industry people, and EPRI experts who benefited from over 20 years of operating experience, including data collection and analysis.

The experience gathered in the nuclear industry has shown that experienced analysts using the PMO process have been able to achieve results similar to the modified classical RCM process, and the concept itself has proven to be highly efficient in terms of achieved savings and overall results [Johnson 1998 and 2003, Harazim & Ferguson 2003, Messaoudi 2005].

As an example of the PMO process advantages, some results concerning improvement of equipment reliability and plant efficiency at Kewaunee Nuclear Power Plant (KNPP) will be presented [Johnson 1998 and 2003]. The plant is situated close to the city of Green Bay (Wisconsin, USA). The PM Optimisation was initiated following the need to shift the plant's PM program from a 12- to 18-month cycle. The technical basis (justification) of maintenance tasks and frequencies needed to be re-established because the original records were not retrievable. Initial efforts to compile similar information consisted of a conventional RCM program. In five years, RCM evaluations were completed on only 8 of the plant's 65 systems. Critical review of the RCM results and progress revealed that although the process was rendering useful information, it was advancing too slowly given KNPP's needs and expectations. In mid-1990s, the KNPP management decided to turn to PMO.

A plan, considering long-term plant operation, was defined with a following main objective: "Maximize plant generating capability and equipment reliability". The other objective was to establish a general plant maintenance program that would minimize power reductions and forced outages resulting from equipment failures, procedural errors, technical inadequacies, and human error. Also, specific objectives were formulated and PMO streamlined approach was used to achieve them. One year after the PMO program had been initiated, the analysis of all plant systems was accomplished. As a result, PM task change recommendations were drawn and, subsequently implemented into existing KNPP programs [Johnson 1998].

An overall evaluation performed four years later [Johnson 2003] revealed that since the beginning of KNPP's Living Program the following significant results have been achieved (only some of them are listed here):

–   Savings resulting from PM interval extension and task elimination over a four year period reached 1.5 M$ (US).
–   Many PM tasks dropped and some new added.
–   Number of PM work orders dropped drastically due to grouping related tasks between plant organisations.
–   12% reduction in maintenance staffing.
–   Reliability Program recognized as strength by US Nuclear Regulatory Commission (USNRC) and INPO.
–   Spin-off of other maintenance processes.
–   Acceptance of a new program was made through staff involvement; etc.

As shown above, the PMO process has proven its efficiency, and is now widely used in the nuclear industry.

In brief, its potential benefits are the following:

–   Elimination of unnecessary PM tasks.
–   Decrease of corrective maintenance in terms of time, man-labour and cost.
–   Reduction of forced outages.
–   Improvement of plant availability.
–   Focus of maintenance resources on critical equipment and components.

PMO balances corrective, preventive and predictive maintenance tasks along with equipment availability and maintenance cost considerations. The ultimate goal consists in maximizing plant value while remaining safe [Liming 2002]. RIAM provides a framework for employing those tools and methods. It extends the scope from plant maintenance policies and practices to all aspects of asset management. Rather than optimising on the basis of reliability, availability and cost, as is the case of a conventional approach, RIAM does an optimisation on the net present value (NPV) as a principal profitability parameter [Liming 2002].

Following encouraging results obtained through PMO applications in the nuclear industry, there are the reasons to believe that its implementation could be rewarding in other industries, too. In this context, the next section discusses some possible applications of the PMO process in mining.

3. POTENTIAL FOR A TRANSFER OF NUCLEAR INDUSTRY KNOW-HOW AND PRACTICES TO MINING OPERATIONS

As mentioned before, conventional RCM is usually a time-consuming process. At the same time, several applications of the ERP and PMO processes in the nuclear industry have delivered encouraging results in terms of saving time, effort and money while remaining safe.

One may expect that these processes may be transferable to the mining industry, particularly in the following circumstances:
– Mills and mineral treatment plants.
– High-scale open-pit mines where capital and operational cost of the equipment as well as its performance are significant contributors to the production cost.
– Existing and future underground mines where production depends (or will depend) to a large extent on highly automated production equipment.

In all cases, in the context of fierce international competition, the mines need the highly effective equipment. Concerning the case of mineral treatment plants, they operate a considerable number of stationary equipment, which makes them somehow similar to nuclear power plants. In this respect, a transfer of know-how and experience could be probably the easiest. In the two remaining cases there are some significant differences between them. Open-pit mines employ huge fleets of capital-intensive equipment and their production systems have got a limited redundancy. Many of them operate around the clock for 365 days a year often-in extreme atmospheric conditions. In these circumstances, ensuring high equipment reliability and availability is a very challenging task and a success depends to a large extent on adequate maintenance policies and strategies. Even if in some cases classical RCM seems to deliver anticipated results, it is worth trying less time- and resource-intensive processes, such as those reviewed in preceding sections.

In what concerns the automated equipment in the underground mines, its operational and service environment is very specific and substantially different from that of man-operated equipment. Due to safety concerns and regulations, zones where automated equipment operates are usually off the limits for human workers. In the case of a failure that requires human intervention, it is an imperative that all man-less traffic and operations in the area concerned be suspended and/or shut down. This may bring a whole section of the mine to a virtual halt and, thus, induce considerable production losses. Also some minor (from a technical point of view) problems that would otherwise be rapidly solved by a human operator may lead to production interruptions. Since in underground mines there is usually even less redundancy (if any) in production systems than at open-pits, automated drill jumbos and/or rigs, LHDs and trucks are really critical for a whole production process.

In each of the above cases, the improvement of reliability and optimisation of proactive maintenance (preventive, predictive, condition based) is a key factor with regard to production efficiency and cost. In this context there is apparently some potential for the application of the ERP and PMO processes in the mining industry.

At the beginning, these processes need to be conceived and developed in the way to adapt them to a specific character and context of mining. At the early stages, the work should focus on some selected categories of equipment considered critical with regard to its role in the production system (including a degree of redundancy), its working conditions and environment, intensity of their use, maintainability and maintenance support. Cable (electric) shovels, hydraulic shovels and crushers in open-pit mines, and underground automated jumbos, production drill rigs and LHD are good examples here. Databases and maintenance templates for those may be developed with an input from OEM, mine operators and maintenance management people, consultants and research institutions.

They are founded reasons to believe that a successful development of specific ERP and PMO and their implementation at mine sites will bring substantial benefits such as:
– Less time and effort deployed, but with similar results compared to conventional RCM.
– Substantial reduction of reactive maintenance, whose share at many mine sites is by far too high.
– Rationalization and optimisation of proactive maintenance tasks will definitely contribute to an increase of availability and reduction of a maintenance cost.
– Better utilization of maintenance resources due to focusing on the truly critical equipment and/ /or its components.

- ERP may be considered almost a closed-loop system that once properly implemented will lay ground for continuous reliability and PM improvement (see Fig. 2).
- Increase of equipment effectiveness and better life-cycle management.

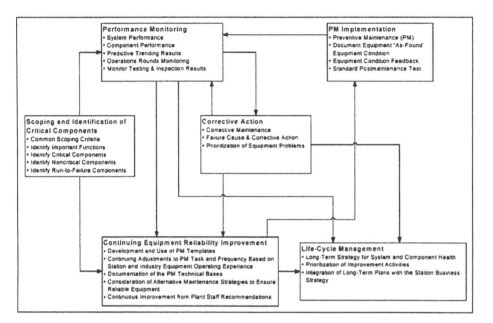

Figure 2. Top Level Diagram of AP-913 Equipment Reliability Process [INPO 2001]

It should not be forgotten however, that the development and implementation of these two processes can never be done instantaneously and their success requires some time, effort, proper understanding, support and commitment as well as patience of all the involved parties (especially from the management).

CONCLUSIONS

Facing tough international competition, mining companies in develop countries are constantly forced to produce more at the lesser cost. One of the most important avenues in this respect is an increase of overall equipment effectiveness. Adequate preventive and proactive maintenance policies and tasks (particularly in what concerns their scope, content and frequency) are among the most crucial factors. However, the mining industry has not yet developed its own original processes for this purpose and it relies predominantly on the know-how and experience gathered in other, more innovative industries. This paper intends to indicate that there is a potential for transferring some important know-how from the nuclear power generation industry. This includes two processes, namely: Equipment Reliability Process (ERP) and Preventive Maintenance Optimisation (PMO). Their successful implementation requires that they be adequately adapted and tailored to fit within the operational and business context of the mining industry. Another indispensable condition is a close collaboration and open-minded exchange between OEM, mine operators and their maintenance management personnel as well as other involved personnel. This approach will be undoubtedly beneficial for mining companies facing a tough competition at the international level.

REFERENCES

Campbell J.D. 1995: Uptime – Strategies for Excellence in Maintenance Management. Productivity Press, Portland, Oregon (USA).

Coppock S. 2004: Raising the Bar, Palo Verde's Secret for Good Equipment Reliability. Nuclear Professional, Atlanta.

DOE-DP-STD-3023-98. 1998: Guidelines for Risk-Based Prioritisation of Doe Activities. (Doe Limited Standard), U.S. Department of Energy AREA MISC, Washington, DC 20585.

Ednie H. 2004: RCM Helps Operations Get the Most out of Equipment, CIM Bulletin, vol. 95, no 1065, pp. 20–25.

Faitakis Y., Mackenzie C. and Powley G.J. 2004: Reducing Maintenance Cost through Predictive Fault Detection. CIM Bulletin, vol. 97, no 1076, pp. 63–66.

Harazim M.L. & Ferguson J.B. 2003: North American Nuclear Maintenance Best Practices Compared to Japanese Utility Maintenance Practices. 24th Annual Conference of the Canadian Nuclear Society, Toronto.

Institute of Nuclear Power Operations (INPO) 2001: AP-913, Revision 1, Equipment Reliability Process Description. Atlanta.

Johnson L.P. 1998: Improving Equipment Reliability and Plant Efficiency through PM Optimisation. Fractal Solutions, Suwanee.

Johnson L.P. 2003: Continuous Improvement with PM Optimisation. Lessons Learned, Fractal Solutions, Suwanee.

Komljenovic D., Vaillancourt R. & Messaoudi D. 2004: Approach Regarding the Implementation of S-98 Reliability Program at the Gentilly-2 Power Plant Considering AP-913 Equipment Reliability Process. 25th Annual Conference of the Canadian Nuclear Society, Toronto.

Knights P.F. 1999: Analysing Breakdowns. Mining Magazine, vol. 181, no 3 (September), pp. 165–171.

Knights P. & Oyaneder P. 2005: Best-In-Class Maintenance Benchmarks in Chilean Open-Pit Mines, CIM Bulletin, vol. 98, no 1088, p. 93 (full text available on the CIM webpage: www.cim.org).

Kumar U. 2003: Economic Aspects of Equipment Breakdowns in Mines and Control Measures. Lecture 5, Division of Mining Equipment Engineering, Lulea University of Technology, Lulea, Sweden.

Kumar U. 2003: Reliability Technique – a Powerful Tool for Mine Operators. Lecture 13; Application of Reliability Centred Maintenance (RCM) Strategy, Lecture 14, Division of Mining Equipment Engineering, Lulea University of Technology, Lulea, Sweden.

Lewis M.W. 2000: Remote-Control Monitoring Cuts Maintenance Cost, Mining Engineering, vol. 52, no 8, pp. 15–19.

Liming K.J. 2002: Risk-Informed Asset Management for Electric Power Plants. Energy Pulse RIAM Paper.

Mercier E., Vallée P., Tapp B., Méthot R. & Clouet J.M. 2004: Maintenance at Falconbridge Ltd.'s Raglan Mine. CIM Bulletin, vol. 97, no 1084, pp. 53–58.

Messaoudi D, Abdul-Nour G., Komljenovic D. & Vaillancourt R. 2003: RCM in Nuclear Power Plants. Canadian Reliability and Maintainability Symposium, Ottawa.

Messaoudi D. 2005: Processus de fiabilité des équipements à la centrale nucléaire Gentilly-2. MSc. Thesis, Université du Québec à Trois-Rivières.

Mitchell J.S. (editor-in-chief) 2002: Physical Asset Management Handbook. 3rd edition, Clarion Technical Publishers.

Moubray J. 1997: Reliability-Centered Maintenance. 2nd edition, New York, Industrial Press Inc.

NUREG-0800. 2002: Chapter 19, Use of Probabilistic Risk Assessment in Plant-Specific, Risk-Informed Decision-making: General Guidance, Revision 1 of Standard Review Plan Chapter 19, U.S. Nuclear Regulatory Commission, Washington, DC 20555-001.

NUREG-1513. 2001: Integrated Safety Analysis Guidance Document, U.S. Nuclear Regulatory Commission, Washington, DC 20555-001.

Paraszczak J., Vachon J. & Grammond L. 1997: Benefits of Studies on LHD Reliability and Availability for Mines. In Strakos et al. (eds), Proc. 6th Int. Symposium on Mine Planning and Equipment Selection, Strakos et al. (eds), Ostrava, Czech Republic, 3–6 September 1997, pp. 469–475, A.A. Balkema.

Turner S. 2002: PM Optimization, Maintenance Analysis of the Future. OMCS Ph 0419397035, Webpage: http://www.pmoptimisation.com.au

International Mining Forum 2006, Sobczyk & Kicki (eds) © 2006 Taylor & Francis Group, London, ISBN 0415 401178

Problems of Underground Ore Mining at Great Depths in Kryvyi Rih Basin

Yuriy Vilkul, Yuriy Kaplenko, Mykola Stupnik, Victor Sydorenko
Kryvyi Rih Technical University, Ukraine

ABSTRACT: The paper presents an analysis of mining methods applied at Kryvyi Rih Iron Ore Complex. Considers the technical and economic indicators of sublevel caving. Offers directions for further development of the technology of ore mining at great depths. Corresponding conclusions are presented.

KEYWORDS: Underground ore mining, naturally rich ores, technical and economic indicators, mining ratio – useful mineral

Further development of underground iron ore mining technologies in Kryvyi Rih Basin – the main iron ore mining area in Ukraine – must be directed first of all at increasing the quality of marketable output, reducing production costs and improving the ecological situation in the region.

Mines that extract naturally rich ores carry out stoping at the depth of 1000–1250 m, development at 1200–1300 m, opening new reserves at 1300–1500 m. Different mining and geological conditions require application of different working methods and their variants. At the same time a specific character of mining in Kryvyi Rih Basin – great depths, high rock stress, relatively soft unstable ores, fractured unstable surrounding rocks – limit the range of methods possible to apply.

Mining and geological conditions do not allow applying highly efficient self-propelled machines and block-caving stoping methods. Most mines limit or even exclude the application of block-and-room methods.

The summaries of mining methods applied at the rich ore mines of Open Joint Stock Company „Krivoy Rog Iron Ore Complex" are given in Table 1.

Table 1. Mining methods applied at Kryvyi Rih Iron Ore Complex

Mines	Block-room [%]	Sublevel-room [%]	Sublevel-caving [%]
"Lenin"	82.1	3.2	14.7
"Gvardeyskaya"	54.6	22.2	23.2
"Oktyabrskaya"	29.0	71.0	–
"Rodina"	–	–	97.8
Total in the Complex	45.0	24.9	29.7

As can be seen from Table 1, the most efficient block-room methods are responsible for less than half of the output. The block-room mining methods provide high labour productivity and least cost per 1 tonne of ore mined. However, under high rock stress, which often occurs in Kryvyi Rih Basin mines their application, is being reduced. Nowadays in mines of the Basin sub-level – room systems are widely used though greater volumes of development workings characterize them. Thus, for block-and-room mining A methods the specific quantity of development workings averages 4.5–6.5 m/1000 t, for sublevel-room methods it fluctuates within the limits of 6.0–8.0 m/1000 t, i.e. 1.3–1.5 times higher. Labour productivity differs correspondingly: 40—60 t/cm for block-level methods and 35–45 t/cm for sublevel-room mining methods.

Sublevel caving is used to mine ores of medium and low hardness and stability under high rock stress conditions. Among variants of the caving method applied in Kryvyi Rih Basin are caving with sets of deep vertical holes into a free-face opening and onto pre-caved ore.

Technical and economic indicators of sublevel caving are given in Table 2.

Table 2. Technical and economic indicators of sublevel caving

| Indicators | Variants | | |
	Caving into a vertical free-face opening ("Kryvorizhstal" Mine)	Caving into a horizontal cut ("Sukhaya Balka" Mine)	Caving onto pre-caved ore ("Rodina" Mine")
Panel area, m^2	500–600	500–600	500–600
Total labour productivity, t/cm	35–45	30–40	25–35
Ore losses, %	15–17	14–15	14–15
Ore dilution, %	4–6	9–11	8–9
Quantity of development workings, m/1000 t	6.2–6.5	6.5–7.5	7.5–8.5

Large specific quantity of development workings makes negative influence on technical and economic indicators of ore mining. This results from the fact that the output per drift miner is 8–10 times smaller than that per driller employed at stoping. Besides, specific power consumption for ore extraction at developing is 4–6 times higher than analogous index at stoping. Nowadays, the cost of cartridge explosives widely used in mining is 3–4 times higher than granulated explosives. Taking into account specific quantity of explosives and their cost, acquisition expenses for explosives used in development are 14–16 times higher than those for stoping. That is why the expenses of preparing a block for stoping make up 32–40% of all the operating costs of mining per block.

The process of formation of a vertical free-face space is technologically complicated and expensive. During the development of a raise the extraction of ore occurs in narrow rock exposure conditions, which decreases hole efficiency parameters and increases the volume of drilling works as well the quantity of explosives for breaking.

The cost involved in the formation of free-face spaces in the operating costs of the ore mining per block varies from 10 to 20%. On the whole, the operating costs per block at "Lenin" and "Rodina" mines vary from 0.8 to 2.00 USD per tonne correspondingly.

The extraction indicators depend to a great extent on the type of deposit and the elements of blocks mined out during the current year.

In general, metal grade loss at the Complex improved by 0.33%.

Actual loss of ore at the Complex have amounted to 9.04 million tonnes over the past 11 years with predicted loss of 9,8 million tonnes. This result was achieved by applying an ore concentration principle enabling to produce ore from blocks with lower Fe content than is possible with the standard method.

The extracted ore is a raw material for the production of marketable iron ore in preparation plants. Kryvyi Rih deposits of hematite ores contain intrusions (up to 6% of volume) of rock with low Fe content. Besides, in the process of mining, dilution of ore with surrounding waste rock takes place. The ores have hardness coefficient of 2–6 and rarely 8–10, whereas waste intrusions have 10–12 and more. That's why the ore is crushed comparatively easily with a cascade of crushers whereas substandard hard rocks can be crushed by means of repeated return to additional crushing in a closed circuit.

The simple technology of ore concentration is based on this specific character of ores. It consists in dividing the ore in the mines into two streams: pure ore and diluted ore. Pure ore is extracted and crushed completely in a closed circuit. Diluted ore is taken out in a separate stream. After crushing according to a closed circuit, the product of screening is marketable sintering ore and oversize fraction +10 mm with Fe content of 44–45% makes the tailings of the concentration process.

Such a technology has enabled to increase the content of iron in sintering ore by 2–4% depending on the volume of intrusions and the degree of dilution. All this has increased competitiveness of the ore. At the same time, the ores of the Complex are still inferior to the ores of other suppliers at the international market.

The operating costs of ore mining depend on the mining and geological factors (thickness of deposit, hardness and water content of ores, quantity of mining, drilling and support work) as well as subjective factors: meeting the planned production volumes, labour management and efficient use of resources. Thus the high operating costs of ore mining at "Rodina" mine are explained by high water hazard (water inflow equals to those in the three remaining mines combined) and by high quantity of development workings and support.

An analysis of expenditure indicates that 5 expenditure items: auxiliary materials; power; wages; maintenance of basic assets; mine general purpose expenditure form 77% of manufacturing cost. So, they are to be the main subject of in-depth study for taking up appropriate steps for cost saving, which is one of the main components in the struggle for the competitiveness of production.

As it was mentioned above the mines of the Complex use the simple concentration principle based on a contrast of strength characteristics of ore and waste rocks. This enables to get quality-sintering ore with increased Fe content due to crushing and sifting. It is necessary to take into consideration the following fact. During the process of ore mining, the dilution rocks with an average Fe content of 36.74% mix with it and the tailings with an average Fe content of 44.75% are dumped during the process of ore concentration. Preliminary calculations show that 2499,4 thousand tons of Fe were mined along with diluted rocks and 4208,6 thousand tons of Fe were lost along with the tailings. This testifies to low efficiency of the concentrating process and calls for urgency in its upgrading. It would be expedient to concentrate ore after each crushing stage. It would enable to increase the grinding efficiency and to decrease power consumption, as power consumption during crushing is 10–15 times higher than power consumption used for screening.

The applied technological methods of producing sintering ore focus on the raw material mined directly in stope blocks.

For this reason, the major problem of underground mining technologies is the increasing of qualitative and quantitative characteristics of extraction. Over the past 11 years the mines of the Complex have lost 9042,8 thousand tonnes of ore with average Fe grade of 55.75% and at the same time 6803,4 thousand tonnes of intrusive rocks with 36.74% Fe content were mined. The annual tonnage of Complex averaged 5957.3 thousand tonnes. An average annual loss comes to 822.1 thousand tonnes. Only 1% in loss reduction enables to market commodity for the sum of U$76.5 thousand and to reduce costs for mining of intrusive rocks with low Fe content.

With regard to the level of 5% losses, the indicators of ore mining and concentration are calculated with fixed crude ore productivity at the Complex (Fig. 1). The given results show that reducing the losses from 13.8% to 5% results in increasing pure ore output and increasing Fe content in it, increasing sintering ore output, decreasing the volume of diluting rock.

The share of individual processes in the mine and site operating costs are given in Figures 2 and 3 correspondingly. Using these data we can conclude that considerable improvement of technical and economic indicators in mines is possible by upgrading the mining methods. It is determined by the fact that the upgrading of underground mining technologies doesn't require capital investments, and there are some reserves in other expense items of operating costs (for example, supplementary materials). Specific quantity of expenditures for supplementary materials in mine operating costs varies from 20.7% ("Rodina" mine) to 28.4% ("Lenin" mine). An aver-age for the Complex is equal to 25.3%.

Development works is the main consumer of mine timber and metal support. Upgrading of mining methods in order to reduce the expenditure on development works will enable reducing the expenditure for the abovementioned supplementary materials. Expenditures on explosives in the Complex amounted to U$1378.2 thousand in 1999, for cartridge explosives U$655.05 thousand. Reducing specific development works cost will allow to purchase expensive cartridge explosives. As the development is the most laborious technological process of underground mining, reducing the volume of development work will enable to increase labour productivity in general.

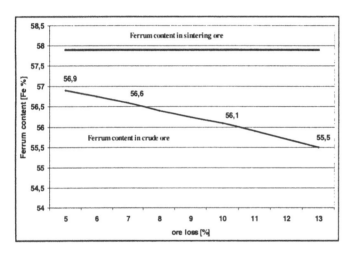

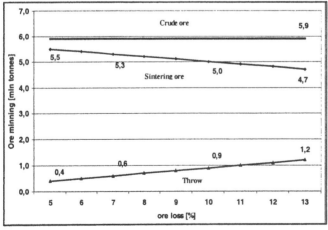

Figure 1. Change of work indicators of the Complex's mines with the reduction of ore loss

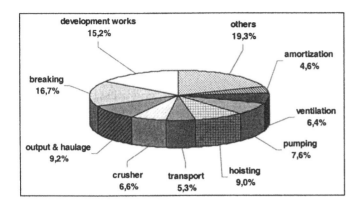

Figure 2. Share of processes in mine operating cost

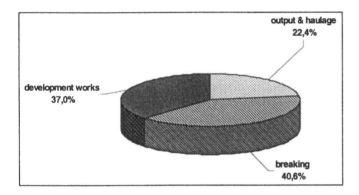

Figure 3. Share of processes in operating cost per block

The characteristic feature of underground mining is the complicated structure of mining me-thods and a large volume of development workings. This results in large amount of manual work to support the workings and large consumption of support materials. Consumption of mine timber is 2.14 m³ per 1,000 tonnes of mined ore, consumption of metal support is 0.585 tonnes per 1,000 tonnes. Foreign enterprises use similar methods, but their distinctive feature is simplicity of method and a sharp decrease in the amount of mining and support works.

The above examples indicate that considerable possibilities for improvement of technical and economic indicators of the underground mining can be achieved by upgrading the mining methods. The key element in increasing the efficiency of extracting minerals is upgrading the actual enginee-ring methods and development of new systems and techniques of underground mining. Along with development works, one of the basic and the most laborious processes of underground mining is ore breaking, which is mostly (74–80%) performed by wells. The specific quantity of labour input in drilling and blasting is 40.6%.

The weak point in the technology of stoping is the process of formation of a free-face breaking space. With a space amount of 25–30%, labour inputs in working the free-face space are equal to labour input in extracting the remaining 70% of the block.

So, the necessity of upgrading the drilling and blasting technique of stoping works is vivid. Moreover, it is relevant not only to the technique of formation of a free-face, but also to the pre-paration of deep wells, as its share in mining cost is 16.7%.

Blasting is a multi-factor process, in which the influence of various factors on the final result isn't the same.

To establish the factors influencing the drilling and blasting efficiency, 9 years' work results of Kryvbass mines were processed. Analysis was done with the random balance method.

The following characteristic (further factors) were processed: the line of least resistance (W), strength of broken ore (f), the specific quantity of explosives consumption for breaking (q), the direction of breaking in respects to layering (α), the mining depth (H), the number of drill-hole rows (N), the volume of broken ore (Q), the type of explosives (Z), the area of free-face rooms (S), and the orientation of the free-face space (Φ).

Estimation of breaking efficiency was done by analysing specific consumption of explosives at secondary breaking.

Diameter of the blastholes was not considered, as it is the same for all the Kryvyi Rih mines. The values of linear and paired factors affecting the quality of ore breaking are given in Table 3.

The studies showed that the depth of pit development, the line of least resistance, physical and mechanical properties of the rock mass, specific consumption of explosives and the value of rock mass exposure are most important, i.e. exert most influence on the quality breaking of rock mass. Thus not only physical and mechanical properties of rock, parameters of blasting and drilling but also mining conditions where rock mass breaking and crushing occur influence the quality of extraction.

Nowadays the share of ore haulage in mine manufacturing cost is 9,2% and is determined by rock size distribution in the blasted rock and loading-hauling equipment type.

Two types of equipment: scraper and vibro-haulage installations are used at the mines of Kryvyi Rih. That is why the real factor providing the increase of haulage efficiency is improving the granulation of the blasted rock mass.

The cost of transforming crude ore into sinter ore is 6.9% of the production cost. The above mentioned data testify of low efficiency of crude ore processing resulting in considerable losses of useful component to dumps, considerable content of −0.25 mm fractions in sinter ore and in increased energy consumption for crushing. This forces the necessity to perfect crude ore processing.

Table 3. Value and grade of basic line paired factors

Grade	Factors	Value, %
1	H	16,6
2	Hf	15,1
3	W	14,5
4	HW	13,4
5	f	11,6
6	Hq	10,7
7	HS	10,2
8	fq	9,4
9	q	8,9
10	Wf	8,8
11	HΦ	7,4
12	fS	7,2
13	S	7,0
14	Wq	6,8
15	HN	6,5
16	WS	6,3
17	qS	5,9

Analysis of technology and technical and economical indicators of underground ore mining in Kryvyi Rih Basin and abroad as well as of the state of iron ore product market in Europe allows to come to the following conclusions:

– Increase of iron ore product quality and technical and economical indicators of ore mining in market conditions is the most important problem of national economy which must be solved in the Complex by perfecting the technological processes of underground ore mining and the ore processing technology at crushing and sorting plants.

– Ore mining cost depends on the extraction ratio, as there exist several cost items whose share in the total cost value is the function of extraction ratio. Prospecting, opening up and deposit development, block cutting and ore extraction costs are these items. That is why considerable cost decrease can be reached by increasing the quality and quantity indicators of recovery and improvement of the named technological processes.

– Useful mineral indicators are determined by the choice of development method, which specifies the expenditure on block development and extraction. So, improving development method with the aim to decrease specific volume of cut workings and the expenditure on creating free-face breaking spaces together with the increase in the quality of recovery and quantity indicators will ensure decreasing the mining cost.

– Development works are the most labour consuming process of underground mining. This is caused by high energy expenditures on rock extraction in conditions of narrow working space, by high cost of manual labour required to support the workings. Thus, for example, expenditure on explosives in development is 14–16 times higher than in production operations taking into consideration energy consumption and the cost of explosives. Besides, the specific expenditures on timber and metal sets are today respectively 214 m^3 and 585 kg per 1000 t of mined ore. That is why the improvement of production indicators can be reached not only by cutting the production expenses but also by improving the drivage methods.

– The bottleneck in the production operation technologies is the process of formation of the free-face breaking space by widening of a cut raise. It requires increased number of shot-holes and higher consumption of explosives for extraction. The expenditures on free-face space formation per block cost change from 10 to 20%.

– The share of expenditure on production of sintering ore in the total mining cost is 6.9% on average at a mine. It is caused by imperfection of the crude ore processing technology in that ore is run through a closed system of crushers leading to over-crushing of its richest part, the increase of energy consumption and high iron losses in the tailings.

International Mining Forum 2006, Sobczyk & Kicki (eds) © 2006 Taylor & Francis Group, London, ISBN 0415 401178

Satellite Remote Sensing as Means to Assess Primary Impact of Open-Pit Mining Development

R. Latifovic
Natural Resources Canada/Earth Sciences Sector. Ottawa, ON, Canada

K. Fytas & J. Paraszczak
Dept. of Mining, Metallurgical and Materials Engineering, Université Laval. Québec City, PQ, Canada

ABSTRACT: Assessment of the possible impact of existing and future mining projects located in remote areas may be difficult and costly. This may have an adverse effect on the process of issuing mine permits and development of new mines. A remote sensing based land cover change assessment methodology presented in this paper is a promising avenue to render this process more meaningful, precise and affordable. It has been applied to a case study of the Oil Sands Mining Development in the north-east of Canadian province of Alberta. So-called Athabasca Oil Sands Region (AOSR) is a place of unprecedented growth of open-pit mining operations. This study was focused on the primary impact onto the landscape and vegetation, considered as all surface disturbances resulting from exploration, mining development, urban development and logging. This impact was assessed using an information extraction method applied to two LANDSAT scenes. The analysis based on derived land cover maps shows a decrease of natural vegetation in the study area (715,094 ha) for 2001 approximately 8.64% relative to 1992. It has been demonstrated that the methodology presented here provides reliable results and is fully applicable for the assessment of the environmental impact of huge-scale mining operations.

1. INTRODUCTION

The mining industry does not have a good track record in what concerns its impact onto environment. Despite the impressive progress in reducing and mitigating its negative effects it is still perceived by many as a harmful, big-scale polluter, leaving behind deep scars in the nature that are difficult and long to heal. This is definitely one of the reasons that the expansion of existing operations or the attempts to open new ones usually encounter considerable reluctance (if not fierce resistance) of the general public and the authorities involved. Understanding of all kinds of interactions between mining activities and the environment is a key factor if a proper balance between the benefits of resource exploitation and its possible adverse effects is sought. Therefore, there is an urgent need to develop and implement tools and methods that would enable the objective assessment of the effects exerted by open-pit mining activities onto surrounding ecosystems.

A successful monitoring approach for evaluating surface processes and their dynamics at a regional scale requires observations with frequent temporal coverage over a long period of time in order to differentiate natural changes from those associated with human activities. However, in some countries such as Canada, mining activities move gradually to remote areas, for which long-term field observations are usually unavailable and/or taking such measurements prior to mine development may not be justified from the economical point of view. In most cases when historical records are needed for studying the long-term vegetation cycles, satellite based remote sensing represents virtually the only alternative to time-consuming and expensive data collection on the ground. This modern technology has already gained widespread recognition as a powerful tool in the studies over different aspects of ecosystems at local, regional and global scales. In addition to being the only available data source in many areas, remote sensing has the added advantage of acquiring data with sufficient area coverage and temporal frequency for studying and monitoring primary impacts caused by surface mining at low cost. It can also be used for studying atmospheric emissions and water pollution indirectly by monitoring green vegetation, an indicator of ecosystem health and conditions. A number of published papers [Ress & Williams 1997, Schmidt & Glaesser 1998, Prakash & Gupta 1998, Wright & Stow 1999] suggest usefulness of such techniques for detecting contamination, determining success in reclaiming open cast mined areas and for providing other relevant spatial data for assessing mining impact on the environment.

Despite its immense potential with regard to environmental monitoring, it seems that satellite based remote sensing has not yet been adequately used in the mining sector. In this context, the principal objective of the study presented here was twofold:

− to provide an initial assessment approach based on remote sensing data, and
− to evaluate the use of such data in investigation of trends and intensity concerning land cover changes in the Athabaska Oil Sands Region (AOSR).

This paper will focus on primary impacts of mining activities onto the landscape and vegetation, considered as all surface disturbances resulting from exploration, mining development, urban development and logging.

2. MATERIALS AND METHODS

2.1. *Study area*

As the world crude oil reserves are steadily getting depleted and the demand for it is on the rise, there has been a growing interest in mining oil-bearing tar sands, whose deposits are abundant in north-eastern part of the Canada's province of Alberta. So-called Athabasca Oil Sands (AOS) deposit, situated north of the city Fort McMurray is one of the world's largest known oil deposits. According to estimates based on exploration results, there are at least 300 billions barrels of recoverable bitumen within 400 km of Fort McMurray (Fig. 1). Two other large oil deposits found also in northeast Alberta are Cold Lake and Pearce River. The AOS deposit has become the site of unprecedented development, with investments in excess of 20 billion US dollars. With several mines already in operation, many more being developed or awaiting statutory approvals, concerns are often raised about the long-term impact they may exert on the surrounding area. The scale, intensity and future perspective of the AOS development have made the region a unique area for studying different aspects of environmental impact. The satellite remote sensing technique offers an additional potential that can be combined with conventional approaches for analyzing spatial and temporal extent of affected areas. Accordingly, this particular region has been chosen as a study area for evaluating a remote sensing based approach for assessment of primary impacts on the surrounding vegetation (Fig. 1).

The study area is situated in the Boreal Plains ecozone, and includes fractions of three ecoregions according to the Canadian Ecological Land Classification System (1995): Wabasca Lowland (ecoregion number 142) in the southwest, Slave River Lowland (ecoregion number 136) in the north and Boreal Uplands (ecoregion number 147) in the west. Cool summers and long cold winters characterize the subhumid mid-boreal ecoclimate type with mean annual temperature of 0.5°C and mean annual precipitation ranging from 350 to 500 mm. Dominant vegetation types are medium to closed canopy stands of aspen (Populus tremuloides) and balsam poplar (Populus balsamifera) with white spruce (Picea glauca), black spruce (Picea mariana) and balsam fir (Abies balsamea). Cold poorly drained fens and bogs are covered with tamarack (Larix laricina) and black spruce. Organic soil is dominant covering 50% of the region. Land uses include forestry, water oriented recreation, wildlife hunting and trapping, as well as oil and gas exploration. Apart from mining and processing of oil sands, other industrial activities are present in the region that may also contribute to stress on the surrounding environment. These include timber harvesting, urban growth of the municipalities of Fort McMurray and Fort McKay, as well as new linear disturbances such as exploration seismic lines, roads and pipelines.

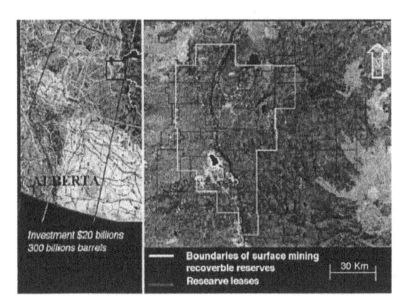

Figure 1. Location of the Athabasca Oil Sands Region (AOSR) study area in the province of Alberta, Canada. The white boundary outlines area of recoverable oil sands reserves

2.2. *Experimental design*

In order to relate surface disturbance and trends in vegetation growth to the proximity of surface mining operations, the study area was spatially delineated into three separate impact zones using potential acid input (PAI) prediction. PAI is the preferred method for evaluating overall effects of acid forming chemicals on the environment since it accounts for the acidifying effect of sulphur and nitrogen species, as well as the neutralizing effect of available base cations. The PAI prediction used in this study is adopted from the Suncor In Situ Oil Sands application. Annual concentration and deposition were predicted using CALPUFF, a multi-layer, multi species non-steady state puff dispersion model which can simulate the effects of time and space varying methodological condition on pollutant transport, transformation and removal (EPA 1995). Three impact zones were defined using isopleths maps representing annual PAI and generic critical load values taken from the Clean Air Strategic Alliance (CASA): low deposition (0.25 keq/ha/year), medium deposition (0.5 keq/ha/year) and high deposition (1.00 keq/ha/year) (Fig. 2). In the following text these areas are referred as Zone-PAI 0.25, Zone-PAI 0.50 and Zone-PAI 0.75 respectively. Direct surface disturbances are assessed for each impact zone by comparing the baseline 1992 versus the current 2001 conditions as defined by land cover maps.

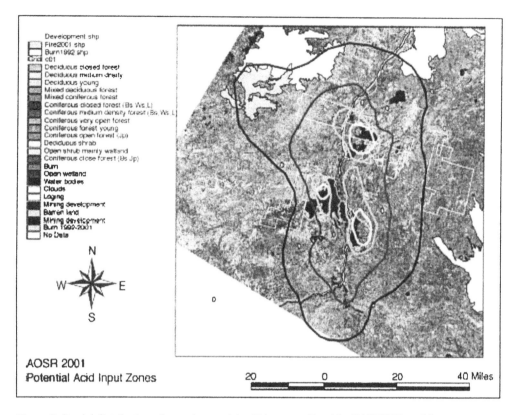

Figure 2. Spatial distribution of annual potential acid input predicted by CALPUFF model run (adopted from in situ Oil Sands Application Firebag project)

3. SATELLITE DATA AND ITS PROCESSING

As mentioned before, in order to properly assess the impact of mining activities onto environment, long-term observations are indispensable. So far, the only available source of long-term terrestrial observations is data collected by several generations of NOAA AVHRR and LANDSAT satellites. The AVHRR (advanced very high resolution radiometer) onboard a series of NOAA satellites have provided daily measurements since 1979 with a coarse spatial resolution of 1.1 km. Despite numerous limitations, the AVHRR historical time series is considered as a primary data set for terrestrial monitoring at regional or global scales. LANDSAT MSS, TM and ETM+, with medium spatial resolutions (30–80 m pixel size) have provided measurements since 1972 at a temporal resolution of 16–18 days.

3.1. Scene selection and study area boundary

Remote sensing based land cover mapping was conducted over the AOS region within the extents of a single LANDSAT scene.

The three images used in this study are:
- LANDSAT 5 TM path: 43, row: 20 acquired on 11 June 1992;
- LANDSAT 7 ETM+ path: 43, row: 20 acquired on 17 August 1999;
- LANDSAT7 ETM+ path: 42, row 20 acquired on 23 August 2001.

The selected images span a 9-year period during the most intensive mining activity.

3.2. Scene geometric rectification

Geometric rectification of the selected images was performed using 20 geographically dispersed ground control points (GCPs) with a first order polynomial and nearest neighbour resampling. An average RMS error below 1.0 pixel was targeted as an acceptable rectification goal. For the 1992 LANDSAT scene an RMS error of 0.65 pixel was achieved. In order to correctly co-register the 1999 and 2001 scenes with the geo-rectified 1992 scene, pixels were identified in each scene corresponding to geo-referenced pixels in the 1992 scene. Rectification of the 1999 and 2001 scenes was performed using these pixels as GCPs. The RMS errors for each geo-rectification were 0.35 pixels and 0.38 for the 1999 and 2001 scene, respectively.

3.3. Scene radiometric normalization

Variation in solar illumination condition, phenology and detector performance results in differences in radiance values unrelated to the reflectance of the land surface. Radiometric normalization was performed using the approach given by Du *et al.* (2001). The process consists of the following steps: (1) selection of pixels pairs in overlap area; (2) principal component analysis for the selection of characteristic pixels; (3) calculation of gain and offset; (4) radiometric normalization.

3.4. Land cover classification

Selected scenes were classified into 15 land cover classes following the method described in Cihlar *et al.* (1998) and Latifovic *et al.* (1999). The images were enhanced using a linear stretch as in Beaubien (1986) to equalize the dynamic range of each spectral band. All pixels in the full-resolution enhanced image were classified into 150 spectral clusters using a K-means classifier. Further cluster agglomeration was performed based on cluster spectral similarity and spatial proximity. The procedure employs basic statistical parameters that determine between cluster spatial proximity and within cluster spectral space data distributions. It uses the arithmetic mean of clusters to compute the Euclidean distance between cluster pairs and the standard deviation to represent clusters' ellipsoid major axis and orientation relative to other clusters. Spectral clusters are first sorted according to decreasing size and then examined for the most similar cluster pairs, starting with the smallest clusters. The spatial criterion is based on the property that many spectral clusters represent gradients within cover types and for these clusters there frequently exists a strong relationship between their spatial and spectral characteristics. The agglomeration procedure utilizes this relationship as a weighting function in computing cluster similarity. More detail on the agglomeration procedure is provided in Latifovic *et al.* (1999). The agglomeration yielded 55 significantly different spectral clusters further grouped into 15 landcover ones (see Table 1). Post-classification refinement included corrections in delineating forest cutblocks, urban and mining development leading to the final land cover maps presented in Figure 3.

Two land cover maps were compared to each other to assess map consistency and then to reference data to assess accuracy. The reference data were project footprints obtained from Environmental Impact Assessment (EIA) Studies of Millennium, Muskeg River and Fort Hills oil sands projects. Other reference data used for the assessment included segments of AlPac Alberta vegetation inventory based on interpretation of 1:20,000 scale aerial photographs from 1991 and series of reports on the Terrestrial Environmental Effect Monitoring (TEEM) Program of the Wood Buffalo Environmental Association (WBEA). Map consistency assessment was performed on a pixel basis over the area that did not undergo change from direct impacts by comparison of pixel label agreement. The assumption was that if both classifications map the same temporally stable pixel with the same label, confidence in mapping accuracy increases. For example, 90% of all pixels mapped as broadleaf forest in the 1992 land cover map were also classified with the same label in the 2001 land cover map. Overall average agreement of all classes weighted by class extent was 87%.

Higher consistency was achieved in mapping forest classes than in mapping non-forest classes (92% and 75% respectively). An analysis of pixels in disagreement showed confusion between spectrally similar or transitional land cover types such as broadleaf forest and broadleaf shrub, treed and wetland or different coniferous sub-classes. Spectrally distinguishable classes, for example, disturb area and forest classes were separated well with an average accuracy of 94%.

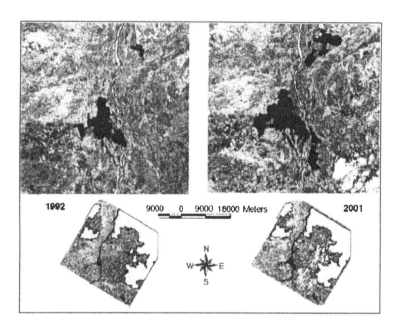

Figure 3. AOSR landscape classifications baseline condition in 1992 and current condition in 2001 derived from LANDSAT TM/ETM+ satellite data

Table 1. Description and hierarchical structure of landscape classes

Landscape class	Description	Label
VEGETATED NATURAL ORIGIN		
DECIDUOUS FOREST		
Deciduous forest	Deciduous trees (aspen, birch, balsam poplar) greater than 60% canopy cover	1
MIXEDWOOD FOREST		
Broadleaf forest	Deciduous and conifer trees: aspen-pine, aspen-white spruce or balsam poplar-white spruce where deciduous fraction is more than 60%. Canopy cover more than 60%	4
Coniferous mixed	Coniferous and deciduous trees where coniferous fraction is more than 60%. Canopy cover more than 60%	5
CONIFEROUS FOREST		
Black spruce forest	Coniferous trees, mainly black spruce with fraction of white spruce and pine. Mosses and shrub dominate in understory. Canopy cover greater than 60%	6
Black spruce young forest	Generally young forest after old perturbation	9
Pine forest	Coniferous trees, mainly jack pine with lichen and some shrub in understory. Canopy cover less than 60%	10
Pine – black spruce forest	Coniferous trees, co-dominant black spruce, jack pine. Mosses, forbs and shrub in understory. Canopy cover more than 60%	13

Landscape class	Description	Label
	NONFOREST	
Treed	Sparse coniferous trees, less than 25% canopy cover. Commonly located on poor sites associated with wetland. This class usually includes a proportion of broadleaf vegetation, mostly shrub	8
Deciduous shrub	Dominated by shrubland or grassland	11
Burn	Burn is usually characterized by sparse vegetation cover varying with the age and intensity of burn	14
Open wetland	Open fen, marsh or swamp	15
	NON VEGETATION	
Barren land	Bare soil, rock space herbaceous and grass	20
Water		16
	ANTHROPOGENIC ORIGIN	
Disturbance	Forestry cutblocks, build-up, roads and exploration lines	18
Industrial development		19
NO DATA	Clouds, shadows	17

4. METEOROLOGICAL DATA

The meteorological data set over the AOSR was extracted from the DS 472.0-TDL US and Canada surface hourly observations generated by the Data Support Section (DSS) of the National Center of Atmospheric Research (NCAR) in Colorado, USA. The complete data set includes measurements of about 1000 meteorological stations across the USA and Canada for the period from 1976 to 2000. In this study, measurements of air temperature at the five stations closest to the AOS region (Table 2) were extracted and analyzed. In addition to NCAR, the metrological data collected by Air Quality Monitoring Stations of the Wood Buffalo Environmental Association (WBEA) for the period from 1998 to 2002 were also included and analyzed.

Table 2. Geographical locations of meteorological stations

Stations	Station ID	Longitude [dd]	Latitude [dd]
Fort McMurray	YMM	-111.22	56.65
High Level	YOJ	-117.17	58.62
Peace River	YPE	-117.43	56.23
Fort Chipewyan	YPY	-111.12	58.77
Slave Lake	YZH	-114.78	55.30

5. CUMULATIVE AND PROJECT SPECIFIC IMPACTS ON LAND COVER DISTRIBUTION. RESULTS AND DISCUSSION

Cumulative and project specific impacts on the landscape are quantified using a post-classification change detection method. This method assumes that reference and compared images are classified to a common legend and that the classification method utilized for landscape mapping provides a high accuracy for both images. Landscape changes are simply detected as differences between pixels' labels. Changes in the landscape distribution in the AOSR for the period 1992–2001 were quantified as the difference between classified images from 1992 and 2001. The assessed areas spatially coincide with the impact zones. The Zone-PAI 0.25 that incorporates the other two zones encompasses 715,094 ha. In 1992 Zone-PAI 0.25 was dominated by native forest vegetation, with natural forest landscape classes occupying 75.6% of the area (Fig. 4 and Tab. 3).

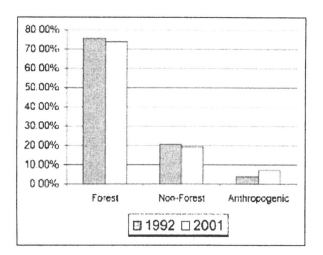

Figure 4. Area occupied by forest, non-forest and anthropogenic landscape classes in AOSR in 1992 and 2001

Table 3: AOSR landscape fraction and class distribution for baseline 1992 and current 2001 conditions

Land cover type	1992			2001		
[ha]	PAI > 0.25	PAI > 0.50	PAI > 0.75	PAI > 0.25	PAI > 0.50	PAI > 0.75
Broadleaf	95,417	39,589	12,639	104,410	32,819	6,934
Coniferous Mixed	35,860	14,521	5,325	39,199	15,333	6,122
Broadleaf Mixed	69,635	27,135	9,475	75,212	29,570	9,769
Coniferous HD	238,475	95,031	47,577	212,751	79,730	38,805
Coniferous Young	58,551	23,193	8,467	62,835	26,679	11,383
Coniferous JP	18,312	3,066	1,705	16,912	2,154	939
Coniferous JP + BS	24,529	6,916	4,156	16,295	5,158	2,564
Coniferous LD	84,855	29,111	14,444	104,587	38,225	17,752
Broadleaf Shrub	24,595	9,609	4,235	9,961	2,857	758
Burn	11,171	481	220	3,447	40	10
Open wetland	1,193	357	171	231	77	47
Water	16,571	8,062	5,034	12,507	5,892	3,882
Barren Land	8,698	2,495	1,225	5,819	2,887	1.294
Logging	6,838	2,645	431	12,920	5,097	709
Mining	20,393	17,482	14,423	38,008	33,177	28,559
SUM	715,094	279,693	129,526	715,094	279,693	129,526
Forest	540,779	209,451	89,343	527,613	191,442	76,516
Non-Forest	147,083	50,115	25,328	136,552	49,977	23,742
Anthropogenic	27,231	20,127	14,854	50,929	38,274	29,268
Forest (%)	75.6%	74.9%	69.0%	73.8%	68.4%	59.17%
Non-Forest (%)	20.6%	17.9%	19.5%	19.1%	17.9%	18.3%
Anthropogenic (%)	3.8%	7.2%	11.5%	7.1%	13.7%	22.6%

Coniferous forest covered approximately half of the study area (48%), whereas deciduous and mixed-wood forest covered 13% and 15% respectively. The remaining 24% of the area consisted of broadleaf shrub, non-forested and open wetland, water bodies, developed land (including active surface mining, roads and built-up), barren land and forest of anthropogenic origin (cutblocks). Two anthropogenic classes together covered 3.8% (or 27,231 ha) of the area. In 2001 Zone-PAI > 0.25 exhibited a similar landscape class distribution as in 1992 but with a slight decrease in the area covered by natural vegetation. From 1992 to 2001, forest-covered area had decreased from 75.6 to 73.8%, while the area covered by natural non-forest vegetation decreased from 20.6 to 19.1%. During the same period, the areas occupied by the two-anthropogenic classes have increased from 3.8 to 7.1% (Table 3). Mining area almost doubled from 20,393 to 38,008 ha and forest cutblocks increased from 6,838 to 12,920 ha.

The spatial distributions of the areas that have undergone change are illustrated in Figs. 5 and 6.

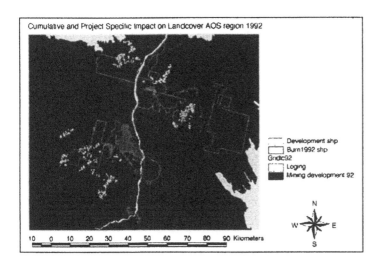

Figure 5. Anthropogenic landscape classes spatial distribution in AOSR in 1992

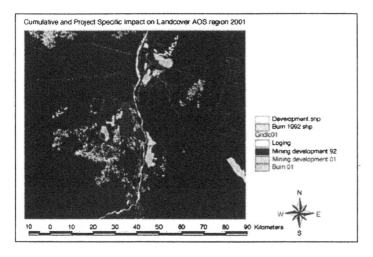

Figure 6. Anthropogenic landscape classes spatial distribution in AOSR in 2001

These figures show an increase in the disturbed area resulting mainly from new developments north and west of the mining development that existed in 1992. An increase in logging activities is also evident, in most cases spatially adjacent to the existing mine site or planned mining developments. However, the amount of logging activity associated with the existence of mining development cannot be evaluated from the data used in this assessment. An analysis of the landscape distribution as a function of proximity to the mining development offers additional insights into the class area changes. Table 3 specifies the landscape class distributions in three surrounding rings at different distances from the mine site. Between 1992 and 2001 the landscape distributions in the Ring-PAI 0.25, the farthest from the mine sites, are more similar than in the Ring-PAI 0.75, which is the closest to the mining development. This difference can be explained mainly by direct impacts due to new surface mining developments in the Ring-PAI 0.75 i.e. the Muskeg River, Millennium, Steepbank, Aurora and North mines, and secondly by logging activities. Areas occupied by the broadleaf and coniferous high-density forest classes in 2001 compared to their respective shares in 1992 have shrunk by 4 and 7%, respectively. Figure 7 shows the changes in forest, water bodies, mining and logging classes relative to 1992 in each ring.

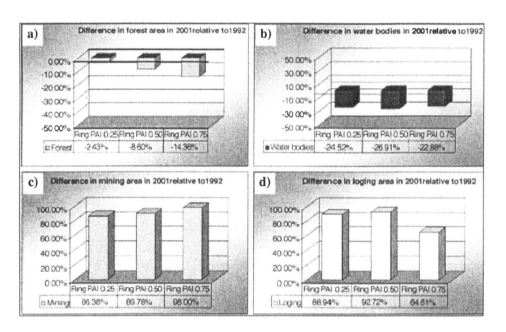

Figure 7. Difference in class area relative to 1992: (a) forest; (b) water bodies; (c) mining; (d) logging

The graph (b) in Figure 7 merits special attention since it exhibits (~25%) decrease in the area of open water bodies. Significant decreases are evident in the vicinity of the Mildred Lake and North mines, suggesting impacts on local hydrology due to these long-term mining operations. Examination of LANDSAT scenes from 1999 and 2002 as well as an ASTER scene from 2000 (15 m spatial resolution) confirmed shrinking of water bodies in the area south west of the Mildred Lake and North mines. These findings need to be confirmed by more detailed analysis of in situ hydrological measurements because the influence on wetland and surrounding vegetation due to a lowering water table could cause significant impacts. The other three graphs in Figure 7 show an overall decrease in natural forest in 2001, an increase in area affected by logging activities for all three rings and a significant increase in disturbed area by surface mining. The results shown on the graph (c) indicate that the area affected by mining in 2001 has doubled in size since 1992. Regardless of relatively high mapping consistency we consider the direct impact presented here as approximate due to the limited capability of medium resolution data to depict small changes less than a pixel size (30 m), such as seismic lines and small clearings for exploration drill holes.

6. CONCLUSIONS

A remote sensing based approach for quantifying primary impacts of surface mining is presented and applied for the assessment of the mining development in Athabasca Oil Sands Region (AOSR). Affected areas were identified as the difference between land cover maps derived from LANDSAT data (30 m resolution) acquired in 1992 and 2001. Maps produced from remote sensing data provide information for subsequent impact assessments from surface mining development on land cover, as well as forming the basis for reclamation planning and monitoring.

Understanding the cause of variability in vegetation conditions allows for distinguishing natural from human-induced perturbations of ecosystems; an important pre-requisite in quantifying environmental impacts caused by mining and other developments. However, it is important to point out that the procedure presented here requires a somewhat complicated procedure for correction and normalization of remote sensing data, including across sensor calibration, atmospheric correction, screening for cloud and other contaminants often present on remote sensing observations. In addition to high quality remote sensing data, a comprehensive knowledge of conditions that a study area was subject to in the past and present is essential for appropriate interpretation leading to higher confidence in obtained results.

REFERENCES

Beaubien J. 1986: Visual Interpretation of Vegetation through Digitally Enhanced Landsat Mss Images. Remote Sens. Rev. 2, 11–43.

Cihlar J., Latifovic R., Chen J.M. & Li Z. 1999: Testing Near-Real Time Detection of Contaminated Pixels in AVHRR Composites. Can. J. Remote Sens. 25, 160–170.

Cihlar J., Hung L. & Xiao Q. 1996: Land Cover Classification with AVHRR Multichannel Composite in Northern Environments. Remote Sens. Environ. 58, 36–51.

Cihlar J., Xiao Q., Beaubien J., Fung K. & Latifovic R. 1998: Classification by Progressive Generalization: a New Automated Methodology for Remote Sensing Multichannel Data. Int. J. Remote Sens. 19, 2685–2704.

Du Y., Cihlar J., Beaubien J. & Latifovic R. 2001: Radiometric Normalization, Compositing and Quality Control for Satellite High Resolution Image Mosaics over Large Areas. IEEE Trans. Geosci. Remote Sens. 39 (3).

Environmental Protection Agency (EPA). 1995. A User's Guide for the CALPUFF Dispersion Model. EPA-454/B-95-006.

Latifovic R., Cihlar J. & Beaubien J. 1999: Clustering Methods for Unsupervised Classification. In: Proceedings of the 21st Canadian Remote Sensing Symposium, II-509-II-515, Ottawa.

Prakash A. & Gupta R.P. 1998: Land-use Mapping and Change Detection in a Coal Mining Area — a Case Study in the Jharia Coalfield, India. Int. J. Remote Sens. 19 (3), 391–410.

Ress W.G. & Williams M. 1997: Monitoring Changes in Land Cover Induced by Atmospheric Pollution in Kola Peninsula, Russia, using Landsat MSS data. Int. J. Remote Sens. 18 (8), 1703–1723.

Schmidt H. & Glaesser C. 1998: Multitemporal Analysis of Satellite Data Their Use in the Monitoring of the Environmental Impact of Open Cast Mining Areas in Eastern Germany. Int. J. Remote Sens. 19 (12), 2245–2260.

Wright P. & Stow R. 1999: Detecting Mining Subsidence From Space. Int. J. Remote Sens. 20 (6), 1183–1188.

International Mining Forum 2006, Sobczyk & Kicki (eds) © 2006 Taylor & Francis Group, London, ISBN 0415 401178

Author Index

Printed and bound by CPI Group (UK) Ltd, Croydon, CR0 4YY

01/11/2024

01782599-0013